E. Y Watson

Hesperiidae Indicae

being a reprint of descriptions of the Hesperiidae....

E. Y Watson

Hesperiidae Indicae
being a reprint of descriptions of the Hesperiidae....

ISBN/EAN: 9783337059828

Printed in Europe, USA, Canada, Australia, Japan

Cover: Foto ©ninafisch / pixelio.de

More available books at **www.hansebooks.com**

HESPERIIDÆ INDICÆ,

BEING A REPRINT OF

DESCRIPTIONS OF THE HESPERIIDÆ,

OF

INDIA, BURMA AND CEYLON,

BY

E. Y. WATSON,
Madras Staff Corps.

MADRAS:
PRINTED AND PUBLISHED BY VEST AND COMPANY, MOUNT ROAD.

1891.

This Book is Dedicated

TO

LIONEL DE NICÉVILLE,

IN PART RECOGNITION

OF

MUCH KINDLY AID RECEIVED.

PREFACE.

THIS book pretends to be nothing more than a collection of descriptions of Indian *Hesperiidæ* which have hitherto been scattered over a large number of periodicals and works which are available only to the favoured few, who have good libraries at their elbow. The original matter in this reprint is practically *nil*, and consists only of the addition of a few localities to those species of which I have a personal knowledge; *viz.*, about one-half of the total number of species described.

The only collections at my disposal are my own private one, and that of the Madras Museum, in which latter *Hesperiidæ* are very poorly represented; however, it is hoped that this book will be found to fill a gap until the publication of Mr. de Nicéville's fifth volume of "The Butterflies of India, Burmah and Ceylon," an event which is hardly likely to take place for some little while.

In addition to the bare descriptions of the several species, I have endeavoured to include all such notes on distribution and synonymy to which I have had access, and have, as much as possible, given authorities for the same. As I have copied all descriptions from every available source, I trust that the original authors will have no objection, since in each case the source from which the description has been obtained is quoted.

The arrangement of the genera I have followed is that of Mr. Distant's "Rhopalocera Malayana," which approximates very closely to that adopted by Mr. Elwes in his recent list of Sikkim Butterflies in the Transactions of the Entomological Society of London for 1888. This arrangement is based on the caudate prolongation or non-prolongation of the posterior wings, and, though perhaps not the most natural, yet is the most convenient to follow as it has been thoroughly worked out, and, in a book of this scope, it would be absurd to attempt any re-arrangement. The arrangement which I should have preferred to follow, and which seems to commend itself as the most natural, is that suggested by Mr. Doherty (Journal of the Asiatic Society of Bengal for 1886) which is founded on the characteristics of the eggs; but this arrangement has not been fully worked out, and Mr. Doherty has lately found it necessary to modify his first tentative sub-divisions.

In the case of synonymy, when there has seemed to be any doubt, I have given descriptions of the two or more species considered by some authorities to be identical, and have given the grounds on which they are supposed to be so. In this family, however, the synonymy is comparatively small, owing chiefly, no doubt, to the fact that until lately comparatively little attention has been paid to the group, and the species in it seem to vary less than in other groups, so that fewer slight varieties have been named.

In all 230 species are described, and as of these over eighty have been discriminated within the last nine years; there is no doubt, however, that numerous other species exist still unnamed.

Mr. de Nicéville has very kindly revised my list of genera and species, and suggested several alterations and additions, all of which, it is needless to say, have been adopted; he has also obtained for me such descriptions as I found inaccessible, and has, at a great expense of time and trouble, revised most of the proofs, and it is difficult to express the thanks that are due to him.

In conclusion, my thanks are especially due to Dr. Edgar Thurston, the Superintendent of the Government Central Museum, Madras, for the free access he has allowed me to the library in his charge, without the use of which this work could never have been attempted.

I had hoped to preface each genus with a short synopsis of species, on the system employed in "The Butterflies of India, Burmah and Ceylon," but have reluctantly been obliged to relinquish the intention owing to the inadequate material at my disposal, and can only hope that the characteristics given with each species will be sufficient to enable it to be identified in nearly every case.

LIST OF THE PRINCIPAL AUTHORITIES CONSULTED.

BUTLER, Catalogue of Diurnal Lepidoptera described by Fabricius in the collection of the British Museum (1869).

CRAMER, Papillons Exotiques (1775—1791).

DISTANT, Rhopalocera Malayana (1882—1886).

DOUBLEDAY, WESTWOOD, and HEWITSON, The Genera of Diurnal Lepidoptera (1846—1852).

DRURY, Illustrations of Natural History (1770—1782).

FELDER, Reise Novara, Lepidoptera (1864—1867).

HEWITSON, Illustrations of New Species of Exotic Butterflies, vol. v (1872—1876).

ID., Descriptions of One Hundred New Species of Hesperidæ (1867, 1868).

ID., Descriptions of New Indian Lepidopterous Insects from the Collection of the late Mr. W. S. Atkinson (1879).

KIRBY, A Synonymic Catalogue of Diurnal Lepidoptera (1871—1877).

KOLLAR, Hügel's Kaschmir, vol. iv (1848).

MOORE, The Lepidoptera of Ceylon (1880-1881).

PROCEEDINGS of the Zoological Society of London (1830 to 1889).

The ANNALS and Magazine of Natural History and Science (1838 to 1889).

The JOURNAL of the Asiatic Society of Bengal (1870 to 1890).

The TRANSACTIONS of the Entomological Society of London (1834 to 1889).

The JOURNAL of the Bombay Natural History Society (1886 to 1890).

LIST OF GENERA AND SPECIES.

I.—BADAMIA.
1. Exclamationis.

II.—UNKANA.
2. Attina.

III.—CHOASPES.
3. Benjamini.
4. Harisa.
5. Anadi.
6. Gomata.
7. Vasutana.
8. Amara.
9. Crawfurdi.

IV.—ISMENE.
10. Œdipodea.
11. Jaina.
12. Mahintha.

V.—HASORA.
13. Badra.
14. Anura.
15. Hadria.
16. Coulteri.

VI.—BIBASIS.
17. Sena.

VII.—PARATA.
18. Chromus.
19. Alexis.
20. Malayana.

VIII.—PADUKA.
21. Lebadea.

IX.—PIRDANA.
22. Rudolphii.

X.—MATAPA.
23. Aria.
24. Druna.
25. Sasivarna.
26. Shalgrama.
27. Subfasciata.

XI.—CAPILA.
28. Jayadeva.

XII.—PISOLA.
29. Zennara.

XIII.—PITHAURIA.
30. Murdava.
31. Stramineipennis.

XIV.—PITHAURIOPSIS.
32. Aitchisoni.

XV.—BAORIS.
33. Oceia.
34. Penicillata.

XVI.—CHAPRA.
35. Mathias.
36. Subochracea.
37. Agna.
38. Prominens.
39. Nostrodamus.
40. Karsana.

XVII.—PARNARA.
41. Guttata.
42. Cingala.
43. Colaca.
44. Bevani.
45. Assamensis.
46. Ornata.
47. Uma.
48. Narooa.
49. Pagana.
50. Plebeia.
51. Kumara.
52. Seriata.
53. Moolata.
54. Canaraica.
55. Austeni.
56. Cahira.
57. Farri.
58. Tulsi.
59. Toona.
60. Eltola.
61. Semamora.
62. Watsonii.
63. Pholus.
64. Sarala.
65. Parca.

XVIII.—SUASTUS.
66. Gremius.
67. Sala.
68. Aditus.
69. Subgrisea.
70. Swerga.

XIX.—SARANGESA.
71. Purendra.
72. Dasahara.
73. Albicilia.

XX.—TELICOTA.
74. Augias.
75. Bambusæ.
76. Siva.
77. Brahma.

XXI.—PADRAONA.
78. Dara.
79. Mæsoides.
80. Pseudomæsa.

81. Gola.
82. Goloides.
83. Palmarum.

XXII.—AMPITTIA.
84. Maro.
85. Coras.

XXIII.—TARACTROCERA.
86. Mævius.
87. Ceramas.
88. Danna.

XXIV.—CUPITHA.
89. Purreea.

XXV.—AEROMACHUS.
90. Indistincta.
91. Obsoleta.
92. Kali.
93. Jhora.
94. Stigmata.

XXVI.—CYCLOPIDES.
95. Subvittatus.

XXVII.—HALPE.
96. Beturia.
97. Sikkima.
98. Separata.
99. Kumara.
100. Aina,
101. Gupta,
102. Cerata.
103. Zema.
104. Dolopia.
105. Radians.
106. Sitala.
107. Honorei.
108. Decorata.
109. Brunnea.
110. Ceylonica.

XXVIII.—ISOTEINON.
111. Atkinsoni.
112. Subtestaceus.
113. Khasianus.
114. Masuriensis.
115. Satwa.
116. Cephala.
117. Cephaloides.
118. Pandita.
119. Flavipennis.
120. Microstictum.
121. Flavalum.
122. Vindhiana.
123. Nilgiriana.
124. Modesta.
125. Flexilis.
126. Masoni.
127. Indrasana.
128. Iapis.

LIST OF GENERA AND SPECIES.

XXIX.—Satarupa.
129 Bhagava.
130 Sambara.
131 Narada.
132 Phisara.
133 Gopala.

XXX.—Calliana.
134 Pieridoides.

XXXI.—Tagiades.
135 Ravi.
136 Khasiana.
137 Distans.
138 Obscurus.
139 Meetana.
140 Alica.
141 Menaka.
142 Atticus.
143 Gana.
144 Minuta.
145 Pralaya.
146 Trichoneura.
147 Helferi.

XXXII.—Abaratha.
148 Ransonnetii.
149 Saraya.
150 Syrichthus.

XXXIII.—Ctenoptilum.
151 Vasava.
152 Multiguttata.

XXXIV.—Odontoptilum.
153 Sura.

XXXV.—Darpa.
154 Hanria.

XXXVI.—Erionota.
155 Thrax.
156 Acroleuca.

XXXVII.—Casyapa.
157 Phanæus.
158 Lidderdali.

XXXVIII.—Gangara.
159 Thyrsis.

XXXIX.—Hidari.
160 Irava.
161 Bhawani.

XL.—Plastingia.
162 Callineura.
163 Noëmi.
164 Naga.
165 Margherita.

XLI.—Hyarotis.
166 Adrastus.

XLII.—Coladenia.
167 Indrani.
168 Tissa.
169 Fatih.
170 Dan.
171 Hamiltonii.

XLIII.—Tapena.
172 Thwaitesi.
173 Agni.
174 Laxmi.
175 Buchananii.

XLIV.—Udaspes.
176 Folus.

XLV.—Notocrypta.
177 Alysos.
178 Paralysos.
179 Albifascia.
180 Restricta.
181 Asmara.
182 Monteithi.
183 Basiflava.
184 Badia.

XLVI.—Celænorrhinus.
185 Pulomaya.
186 Flavocincta.
187 Pyrrha.
188 Plagifera.
189 Patula.
190 Pero.
191 Sumitra.
192 Leucocera.
193 Putra.
194 Munda.
195 Ambareesa
196 Chamunda.
197 Nigricans.
198 Fusca.
199 Spilothyrus,
200 Consertus.
201 Cacus.
202 Tabrica.
203 Pinwilli.

XLVII.—Hantana.
204 Infernus.

XLVIII.—Astictopterus.
205 Xanites.
206 Butleri.
207 Olivascens.
208 Salsala.
209 Subfasciatus.

XLIX.—Kerana.
210 Diocles.
211 Aurivittata.
212 Gemmifer,
213 Dhanada.

L.—Baracus.
214 Vittatus.
215 Subditus.
216 Septentrionum.

LI.—Hesperia.
217 Dravira.
218 Marrubii.
219 Cashmirensis.
220 Galba.
221 Evanidus.
222 Zebra.

LII.—Lobocla.
223 Liliana.
224 Casyapa.

LIII.—Gomalia.
225 Albofasciata.
226 Litoralis.

Species Incertæ Sedis.
227 Avanti.
228 Dimila.
229 Cyrina.
230 Decoratus.

INTRODUCTION.

THE HESPERIIDÆ comprise the fifth Family of the order Rhopalocera as generally subdivided. They are characterized as follows in the "Butterflies of India, Burmah and Ceylon":—

"HESPERIIDÆ, with all six legs perfect. Wings, with the discoidal cell of hindwing slenderly, and often incompletely closed; subcostal nervure of forewing always with four branches, all four emitted before end of cell. Of small size, very robust build and rapid flight. Body of all but a few very robust; antennæ, wide apart at base, with a thick club, or strong curved hook at tip; palpi, short, very broad, closely pressed against the face, densely squamose. Hindlegs, generally with a pair of movable spines at the tip of the tibiæ, and another pair in the middle; middle legs with a pair of movable spines at the tip of the tibiæ."

The eggs of this family have been studied by Mr. Doherty, and he has proposed* a classification of the Indian genera founded on the characteristics of the egg.

This proposed classification is quoted at full.

"HESPERIIDÆ.—Eggs, very large, very few (except in the first group), only one or two matured at a time; opaque, dome-shaped, smooth; or with delicate, depressed serrate ribs, few or very numerous, and with distinct cross lines.

SUB-DIVISIONS OF THE *HESPERIIDÆ.*

HESPERIINÆ.—Egg, small, hard, seven-eighths as high as wide or even higher, constricted at base, with wide, scalloped, anastomosing ribs. This group is very distinct. The only Indian genera belonging to it are *Hesperia* (*Pyrgus*) and *Gomalia*.

SUASTINÆ.—Egg, lower, dome-shaped, large, hard, constricted at base, with a few broad depressed, delicate, biserrate ribs. This group does not seem to be represented in Europe or North America.

BAORINÆ.—*Cyclopides* Group.—Egg, similar, two-thirds as high as wide, constricted at base, with very numerous slender ribs.

BAORINÆ.—*Baoris* Group.—Egg, half as high as wide, leathery, limpet-shaped, widest, and often carinate at base, smooth, generally overlaid with pigment above, as in many *Papilios*, sometimes with numerous obsolete ribs. This group seems to be equivalent to the *Astyci* as defined by Scudder."

Under *Suastinæ* Mr. Doherty includes *Badamia, Choaspes, Plastingia, Suastus, Hyarotis, Tagiades, Sarangesa, Satarupa, Abaratha, Antigonus, Halpe, Isoteinon,* and *Coladenia*; under *Baorinæ* (*Cyclopides* group), *Cyclopides* and part of the old genus *Plesioneura*; under *Baorinæ* (*Baoris* group) remainder

---

*J. A. S. B., Vol. LV, pt. 2, 1886, p. 111.

of *Plesioneura, Udaspes, Parnara, Chapra, Thanaos, Taractrocera, Telicota,* and *Padraona.*

Mr. Doherty also remarks that " a kind of hermaphroditism seems to occur sometimes in the *Hesperiidæ.* From the body of (apparent) males of *Suastus eltola* and of *Coladenia dan,* both having perfect prehensores of the form characteristic of their respective species, I obtained one or two well-developed eggs exactly similar to those taken from the females of the same species. Also, from a male of *Suastus toona* (the egg of that species being, except for this, unknown to me) I obtained a single immature blood-red egg. I have not observed this in any of the higher groups of butterflies."*

In connection with the above it may be noted that the two species *eltola* and *toona,* which Mr. Doherty includes under *Suastus,* are now generally included under *Parnara,* which genus is included by Mr. Doherty under a different sub-division to that of *Suastus,* so either the egg classification is unreliable, or these two species should be kept distinct from *Parnara.*

In the larvæ the head is always large and the neck narrow. The pupæ are more moth-like than those of any other family of butterflies, and are generally enclosed in a web between two leaves.

The perfect insects are readily distinguished on the wing by their short jerky flight, which has earned for them the name of " skippers." When at rest the wings are either raised above the back or else extended flat, in the latter case frequently on the underside of a leaf, and in some genera the forewings are sometimes raised above the back and the hindwings depressed. Many genera are markedly crepuscular in the perfect state.

As it may possibly assist in their classification, I append lists of the genera which rest with their wings raised and extended respectively.

With their wings raised: *Badamia, Choaspes, Ismene, Hasora, Parata, Pithauria, Baoris, Suastus, Ampittia, Cupitha, Thanaos, Cyclopides, Erionota, Halpe, Taractrocera, Isoteinon, Udaspes, Notocrypta, Gangara, Hyarotis, Astictopterus, Baracus, Gomalia* and *Matapa.*

With their wings extended flat: *Sarangesa, Satarupa, Tagiades, Abaratha, Coladenia, Tapena, Lobocla, Celænorrhinus* and *Calliana.*

The following genera rest with the wings either closed over the back, or with the forewings raised above the back and the hindwings extended flat: *Chapra, Parnara, Telicota, Padraona,* and *Ampittia.*

I have no information or personal experience of the genera *Pirdana, Bibasis, Capila, Pisola, Pithauriopsis, Plastingia,* and *Hantana.* In the genus *Hesperia* I have experience of three species, *viz., H. galba, H. evanidus* and *H. marrubii;* of these the two former rest with their wings raised, and the latter (which should probably not be included in the same genus) with its wings extended.

---

* Mr. Doherty has subsequently stated (J. A. S. B., Vol. LVIII, pt. 2, 1889, p. 126, foot-note) that he has found it necessary to altogether remodel his arrangement of the *Hesperiidæ,* but this modification has not yet been published.

HESPERIIDÆ.

GENUS I.—BADAMIA.

Hesperia, Butler, Ent. Month. Mag., 1870, p. 58.
Badamia, Moore, Lep. Cey., vol. i., p. 157, (1881).

" Forewing, narrow, elongated; costa slightly arched at base, exterior margin very oblique and slightly convex below the apex; cell very long and narrow, extending three-fourths the wing; first subcostal branch emitted at two-fifths, second at one-fifth, third at one-seventh, fourth close to and fifth at end of the cell; disco-cellulars very slender, inwardly oblique, of nearly equal length, upper bent inward close to subcostal; upper radial from the angle, lower from their middle; median branches curved at their base, middle branch emitted at about one-fourth, and lower at three-fourths before end of the cell; submedian curved in the middle; hindwing short; apex very convex, angularly lobed at anal angle, abdominal margin short; præcostal projecting inward; costal vein arched upward from the base; second subcostal emitted at one-third from the base; cell broad throughout; disco-cellulars very slender, scarcely visible, of equal length; radial from their angle, very slender; middle median at about one-third, and lower at one-fifth from the base; submedian straight, internal slightly curved. Thorax stout; abdomen rather long, attenuated; head broad; palpi broad and flattened in front, bristly on outer edge, third joint long, projected forward, cylindrical; fore-tibiæ tufted beneath, femora slightly pilose beneath; antennæ with a lengthened club and long pointed tip.

Type *H. exclamationis.*" (*Moore, l.c.*)

1.—BADAMIA EXCLAMATIONIS, *FABRICIUS.*

Papilio exclamationis, Fabricius, Syst. Ent., p. 530 (1775).
Papilio ladon, Cramer, Pap. Exot., iii, pl. 284, fig. C. ♀.
Badamia exclamationis, Moore, Lep. Cey., vol. i, p. 157, pl. 66, figs. 2, a. b. (1881).

" Upperside dark purplish brown, the base of both wings greyish olive brown. Male. Forewing with three transparent slender yellow spots disposed longitudinally on the upper disc, the inner spot ending within the cell. Female. Forewing with the spots larger, the middle spot oblique and irregularly angulated; a less distinct spot also above the middle of submedian vein. Underside pale greyish brown: forewing with discal area darker brown, the spots as above, and pale ochreous posterior border: hindwing with a dark brown anal area bordered above by a short pale ochreous streak. Thorax greyish olive brown; abdomen dark brown with pale ochreous segmental bands; head and palpi in front pale ochreous with brown streaks; third joint of palpi brown; legs brown above, pale beneath.

Expanse $1\frac{3}{4}$ to $2\frac{1}{4}$ inches.

Larva pale violaceous yellow, with numerous black transverse dorsal lines; stigmata whitish encircled with black. Feeds on Terminalia. Pupa violaceous." (*Moore, l. c.*)

Recorded from Ceylon, Sikkim, Nicobars, Andamans, Calcutta, Kumaon, Cachar, Kangra, Bombay, Poona, Belgaum, Nilgiris, and Orissa.

A common species everywhere.

According to Mr. Elwes this species in Sikkim varies considerably in size, and in the number and size of the spots on the forewing.

I have this species from the Deccan, Ganjam, Mysore and Madras; the spots on the forewing of the male are sometimes obsolete.

Mr. Butler (Ann. and Mag. of Nat. Hist., sixth series, vol. i, p. 206) is inclined to consider *B. ladon* as distinct from *B. exclamationis*, but as the former is universally admitted to be the female of the latter, and, as Mr. Butler himself apparently considered the two species as identical in his catalogue of the Fabrician Lepidoptera, it would seem highly improbable that the two names apply to two distinct species, apparently the only difference between the two forms lies in the number of spots on the forewing, a character which is eminently variable.

In collections Indian Museum and de Nicéville.

GENUS II.—UNKANA.

Unkana, Distant, Rhop. Mal., p. 370 (1882-86).

"Anterior wings elongate; costal margin moderately convex, outer margin oblique, inner margin nearly straight, a little shorter than outer margin. Costal nervure extending to about half the length of costal margin; first subcostal nervule emitted at about equal distances apart between base of first and end of cell; fifth from end of cell: disco-cellular nervules obliquely directed inwardly, the upper distinctly longer than the lower; second median nervule emitted much nearer upper than lower median nervule. Posterior wings elongate and somewhat lobately produced near anal angle, the outer margin obliquely convex. Subcostal nervules bifurcating beyond middle of cell; second median nervule emitted nearer to upper than lower median nervule. Body robust; palpi broad and pilose; antennæ moderately long, their apices incrassated, with the tip attenuated and curved or hooked; legs long, anterior tibiæ short and thickened; posterior tibiæ with two long and prominent spines near apex.

Unkana is allied to *Badamia*, Moore, and includes three species which are at present known as found in this (*i.e.*, Malayan) fauna." (*Distant, l.c.*)

2.—UNKANA ATTINA, *HEWITSON.*

Hesperia attina, Hewitson, Trans. Ent. Soc., third series, vol. ii, p. 489, n. 10 (1866).

Hesperia attina, Plötz, Stett. Ent. Zeit., vol. xliii, p. 339, n. 122 (1882).

Hesperia latreillei, Felder, Reise Nov., Lep., vol. iii, p. 511, n. 892, t. 71, fig. 8 (1866).

Unkana attina, Distant, Rhop. Mal., p. 371, pl. xxxiv, fig. 30 (1882-86).

"Wings above dark fuliginous; anterior wings with eight pale irregular spots, of which four are in oblique series from upper discoidal nervule to lower median nervule, three divided by the fourth and fifth subcostal nervules and one in cell; posterior wings with the disk pale greyish. Wings beneath somewhat paler than above; anterior wings spotted as above, and with some submarginal greyish streaks near apex; posterior wings with the pale discal area much larger than above, and extending to the outer margin between apex and median nervules. Body more or less concolorous with wings.

Exp. wings, 55 millim.

Habitat, India *sic.* (*Hewitson*); Malay Peninsula, Malacca (*Biggs, coll. Distant*); Java (*Felder and Hewitson*)." (*Distant, l. c.*)

In collection de Nicéville.

GENUS III.—CHOASPES.

Choaspes, Moore, Lep. Cey., vol. i, p. 158 (1881).
Choaspes, Distant, Rhop. Mal., p. 372 (1882-86).

"Forewing somewhat narrower and more regularly triangular than in *Ismene*; cell broader; first subcostal emitted at one-third before end of the cell; disco-cellulars very oblique; middle median emitted at one-fourth and lower at three-fourths before end of the cell. No glandular patch in male; hindwing somewhat shorter; anterior margin more oblique, and margin prominently lobed, abdominal margin longer; second subcostal emitted at one-third from the base; middle median from near end of the cell, lower at one-half before the end; antennæ much slenderer throughout.

Type, *C. benjamini.*" (*Moore, l. c.*)

3.—CHOASPES BENJAMINI, *GUÉRIN.*

Thymele benjamini, Guérin, Delessert's Souv. Voy. Ind., ii, p. 79, pl. 22, figs. 2. 2 a (1843).

Hesperia xanthopogon, Kollar, Hügel's Kaschmir, vol. iv., p. 453, pl. 18, figs. 1, 2. (1844).

Choaspes benjamini, Moore, Lep. Cey., vol. i, p. 159, pl. 64, figs. 1. a. b. (1880-81).

"Male. Upperside glossy bluish purple olive brown, the basal area more distinctly olive brown. Cilia of hindwing and anal lobe broadly ochreous-red. Female, deeper olive brown. Underside glossy aënescent olive brown, the veins black; forewing with a broad pale cupreous brown band on posterior margin; hindwing with a broad ochreous-red lobular patch with black macular upper border and broad central angular streak. Thorax greyish olive above, vertex bluish olive, abdomen brown; palpi and thorax in front and abdomen beneath, ochreous-red.

Expanse, 2 to 2¼ inches.

Larva with broad transverse dorsal black and yellow bands and two rows of white spots along the back; head, two anal segments and laterally below the bands red; face black spotted. Pupa pinkish grey black spotted.

Central Provinces [Ceylon]. Hills 3,000 to 5,000 feet. Forests. May to November. Shy, but not difficult to capture. Settles on ground; easily disturbed but returns to same place *(Hutchison)*.

Dickoya [Ceylon]. Not common. The larva rolls itself upon the tip of the leaf on which it feeds, and when it has eaten this leaf it goes to another, and so on till it changes to pupa *(Wade).*" *(Moore, l. c.)*

Also recorded from Kumaon *(Doherty)*; Cachar *(Wood-Mason and de Nicéville)*; Nilgiris *(Hampson)*; Ahsown, N. Tenasserim *(Limborg)*; Kangra *(Moore)*; Sikkim *(Elwes)*, Sikkim *(de Nicéville)*.

This is a species which varies very little. I have it commonly from the Nilgiris.

In collections Indian Museum and de Nicéville.

4.—CHOASPES HARISA, *MOORE.*

Ismene harisa, Moore, P. Z. S., 1865, p. 782.

Choaspes harisa, Distant, Rhop. Mal., p. 373, pl. xxxiv, fig. 22 ♂. (1882-86).

Choaspes harisa, de Nicéville, J. A. S. B., vol. lii, pt. 2, 1883, p. 84, pl. 10, fig. 8 ♂.

" Male and female brown.

Male. Upperside dull vinaceous brown, palest on the disk; forewing with an orange yellow costal streak; hindwing broadly along anterior margin pale buff yellow. Body greyish. Cilia of hindwing orange yellow. Underside paler suffused with orange yellow; forewing with a curved series of pale purple narrow streaks between the veins before the apex, and a broad pale buff patch along the posterior margin; hindwing with the veins and lines between them and cilia orange yellow: a black orange yellow encircled basal spot on both wings; a discal series of pale purplish streaks. Third joint of palpi brown; palpi beneath, front and sides of thorax, legs and streak along side of abdomen orange yellow; middle of thorax and abdomen and anal tuft orange yellow.

Female. Upperside dark purple brown; the base of wings greyish, with steel blue gloss. Body greyish. Cilia of hindwing pale orange yellow. Underside as in male; posterior margin of forewing with a less defined pale patch.

Expanse, 2 inches.

Habitat, Darjeeling." *(Moore, l. c.)*

Also recorded from the Andamans *(Wood-Mason and de Nicéville)*; Sikkim *(de Nicéville)*; Sikkim *(Elwes)*.

I have obtained this species commonly in Rangoon.

In collections Indian Museum and de Nicéville.

Mr. Elwes describes a single fresh female of this species in his collection from Sikkim as blackish brown, brilliantly glossed over with steely green which extends nearly to the border of the hindwing and over more than half the forewing. It lacks the costal streak of the male.

5.—CHOASPES ANADI, *DE NICÉVILLE.*

Choaspes anadi, de Nicéville, J. A. S. B., vol. lii, pt. 2, 1883, p. 83, pl. 10, fig. 6 ♂.

"Male. Upperside dark vinaceous brown distinctly glossed with purple, slightly paler in the middle of the disc. Forewing with a costal streak from the base to beyond the middle of the wing rich orange; cilia cinereous. Hindwing with the costa broadly pale ochreous; the cilia rich orange. Base of both wings and thorax clothed with long pale green iridescent hairs. Underside paler brown washed with ochreous, which colour assumes indistinct streaks between the veins on the hindwing. Forewing with the outer margin broadly washed with deep purple, the inner margin broadly pale ochreous; some pale streaks between the veins beyond the end of the cell; a round black spot at the extreme base of the wing with a spot of bright orange above it; hindwing with a similar but larger black spot. Antennæ dark brown above, ochreous below; palpi with the third joint dark brown, the second and first with the outer edge brown, the rest orange, which is the colour of the legs, the underside of the body and the anal tuft.

The female differs from the male only in being larger and darker, the upperside of the hindwing concolorous with the rest of the wing, not broadly pale ochreous as in the male.

The male of this species closely resembles that sex of *C. harisa*, Moore, but differs in the forewing being much narrower, and on the hindwing in having the costal pale patch more restricted; on the underside the markings are less prominent.

There is a male of *C. anadi* from Masuri, taken at 7,000 feet elevation on 27th May, 1868, in Colonel Lang's collection. Expanse: ♂ 1·9 to 2·1; ♀ 2·45 inches.

Habitat, Sikkim; Masuri." (*de Nicéville, l. c.*)

In collections Indian Museum and de Nicéville.

6.—CHOASPES GOMATA, *MOORE.*

Ismene gomata, Moore, P. Z. S., 1865, p. 783, ♂.

Choaspes gomata, de Nicéville, J. A. S. B., vol. lii, pt. 2, 1883, p. 83, ♀, pl. 10, fig. 7.

"Male. Upperside pale vinaceous brown; both wings with pale brownish yellow streaks longitudinally between the veins. Abdomen blackish brown with yellowish bands. Cilia yellowish. Underside dark brown, with the veins and longitudinal streaks between them greyish green, the brown showing only along each side of the veins; posterior margin of forewing broadly pale vinaceous; exterior margin of both wings defined by a brown line. Third joint of palpi and edge of sides brown, the rest yellow. Thorax, legs and abdomen beneath orange yellow.

Expanse, 2¼ inches.

Habitat, N. E. Bengal." (*Moore, l. c.*)

"Female. Expanse 2·3 inches. Upperside very dark glossy bronzy-green, shading off into glossy indigo-blue at the apex and outer margin. Underside with the markings and ground-colour darker than in Sikkim males; forewing with a pale green spot in the second median interspace, with a larger one in the interspace below it, in the male these spots are merged in a large patch of the ochreous ground-colour from the inner margin. The green markings everywhere more restricted and of a darker shade than in the male.

The specimen figured, taken by Mr. Rhodes-Morgan in the Wynaad, is the only female I have seen; there are numerous males, however, in Mr. Otto Möller's collection from Sikkim." (*de Nicéville, l. c.*)

Also recorded from the Nilgiris (*Hampson*); Sikkim (*Elwes*).

In collections Indian Museum and de Nicéville.

7.—CHOASPES VASUTANA, *MOORE*.

Ismene vasutana, Moore, P. Z. S., 1865, p. 782.

"Male. Upperside deep purple brown, paler on the base of the wings; forewing with orange yellow costal basal streak. Cilia of hindwing broad and bright orange yellow. Underside glossy greyish green, the veins and narrow intermediate parallel lines blackish; a patch on posterior half of forewing brown bordered above with blue. Third joint of palpi brown, the rest orange yellow. Head, thorax in front and beneath, legs, middle of abdomen beneath, and anal tuft bright orange yellow.

Female. Upperside darker brown, the base of the wings greyish blue; forewing with two small semitransparent spots obliquely beneath the extremity of the cell. Underside as in male, with the two spots as above.

Expanse, ♂ 2¼, ♀ 2¼ inches.

Habitat, Darjeeling." (*Moore, l. c.*)

Also recorded from Sikkim by Mr. de Nicéville and Mr. Elwes, and Mr. de Nicéville informs me it is very common in the Khasi hills.

In collections Indian Museum and de Nicéville.

8.—CHOASPES AMARA, *MOORE*.

Ismene amara, Moore, P. Z. S., 1865, p. 783.

"Male and female. Upperside brown with a greenish gloss; costal streak of forewing ochreous yellow in the male, less prominent in the female; male with a blackish subbasal patch. Cilia of both wings short and brownish white. Body dark brown; abdomen with greyish segmental bands. Underside, forewing brown, becoming bluish black along the base of the costa; posterior margin broadly brownish white; hindwing bluish black; veins of both wings brownish white, the space between them having a greyish blue parallel line running their entire length. Both wings also with the black ochreous-yellow-encirled basal spot. Thorax in front and beneath, head, palpi, legs, middle of abdomen, and anal tuft ochreous yellow. Femora and tibiæ with a black spot; sides of abdomen black, the segmental bands prominent. Cilia greyish.

Expanse, 2½ inches.
Habitat, N. E. Bengal." (*Moore, l. c.*)
Also recorded from Sikkim by Messrs. Elwes and de Nicéville.
In collections Indian Museum and de Nicéville.

9.—CHOASPES CRAWFURDI, *DISTANT*.

Ismene crawfurdi, Distant, Ann. and Mag. Nat. Hist., fifth series, vol. x, p. 247 (1882).

Choaspes crawfurdi, Distant, Rhop. Mal., p. 372, pl. xxxiv, fig. 26. (1882-86).

" Wings above obscure olivaceous green, becoming tinged with fuscous towards outer margins. Posterior wings with a large anal angular bright yellowish patch, inwardly and broadly margined with black, apical portion of abdominal margin also bright yellowish. Wings beneath paler and more metallic green, the nervures and nervules distinctly darker ; posterior wings with a very large bright yellowish anal angular patch, which extends from about middle of abdominal margin to between the second and third median nervules, and which possesses a long black irregular streak on inner side of submedian nervure, two parallel black streaks between the submedian nervure and lower median nervule, and two similarly placed black spots between the second and third median nervules on outer edge of the yellow patch. Body above more or less concolorous with wings, anal tuft yellow ; body beneath and legs paler.

Expanse, 52 to 58 millims.

Habitat, Malay Peninsula ; Penang (*Biggs—coll. Distant*) ; Province Wellesley (*coll. Distant*)." (*Distant, l. c.*)

There is a single specimen of this species in the Rangoon Museum obtained at Myitta, Tavoy. There appears to be no other record of this species within British limits.

GENUS IV.—*ISMENE*.

Ismene, Swainson, Zool. Illust., vol. i, pl. 16 (1820-21).
Ismene, Moore, Lep. Cey., vol. i, p. 157 (1881).

" Forewing short, broad, triangular; costa much arched at the base, exterior margin oblique, shorter than posterior margin ; cell extending to more than half the wing, very narrow at the base ; first subcostal emitted at beyond one-third, second and third at equal distances between first and end of the cell, fourth at half distance between third and fifth ; disco-cellulars inwardly oblique, of equal length, upper bent inward near the subcostal ; upper radial from its angle, lower from their middle ; upper median branch arched at the base within the cell, middle median at two-thirds before end of the cell and close to the lower median ; submedian recurved. Male with a large basal circular glandular patch of raised scales : hindwing short, broad, anterior margin much arched, exterior margin convex, abdominal margin short ; costal vein arched towards the end, cell short ; second subcostal at nearly one-third from the base ; disco-cellulars very slender, outwardly oblique, concave; radial from their middle; two upper median branches from end of the cell, lower from one-third before the end; submedian and internal slightly curved. Body very robust; palpi thick, flat in front, bristly at the side, third joint long, naked, cylindrical; antennæ short, thick; tip long and slender. Femora pilose beneath; tibiæ tufted; hind tibiæ also incrassated.

Type, *I. œdipodea.*" (*Moore, l. c.*)

10.—ISMENE ŒDIPODEA, *SWAINSON*.

Ismene œdipodea, Swainson, Zool. Illust., vol. i, pl. 16 (1820-21).

Ismene œdipodea, Moore, Lep. Cey., vol. i, p. 158, pl. 64, figs. 2. a.b. (1881).

"Male. Upperside ochreous olive brown; forewing with an ochreous-red costal band which also extends across base of the cell; a large black basal patch below the cell. Cilia of forewing pale brownish grey, of hindwing ochreous-red. Female differs above only in the absence of the basal black patch, and beneath it in the less prominent white posterior marginal band. Underside ochreous-brown: forewing with a paler ochreous subapical and a marginal fascia, and a broad whitish posterior band: hindwing with bright ochreous red longitudinal streak between the veins, broadest between the median and submedian veins and abdominal margin; a small black spot at the base above the costal vein. Thorax in front, head, palpi, body beneath, and legs ochreous-red; terminal joint of palpi brown.

Expanse, ♂ 2, ♀ 2¼ inches.

Near Trincomallee; Kandy; Balangada [Ceylon]." (*Moore, l. c.*)

Also recorded from the N.-W. Himalayas (*Moore*), Cachar (*Wood-Mason and de Nicéville*), Sikkim (*Elwes*).

Mr. Elwes notes that a male from Sikkim differs from two from Kangra in having the costal margin of the hindwing above distinctly white. The black velvety patch is very distinct, whereas it is very faint in *I. jaina*. First described from Sumatra.

The female also differs from that sex of *I. jaina* in possessing the red costal streak of the male which is wanting in the female of the latter species.

In collections Indian Museum and de Nicéville.

11.—ISMENE JAINA, *MOORE*.

Ismene jaina, Moore, P. Z. S., 1865, p. 782.

"Male and female dark vinaceous brown.

Male. Upperside—forewing with an orange red subcostal basal streak and an indistinct blackish patch beneath the cell; front of thorax, anal tuft, and cilia of hindwing bright orange red; thorax and base of abdomen clothed with bluish grey hairs. Underside paler brown; both wings with a small black orange red bordered basal spot: forewing with a well defined purplish white spot within the cell; and a curved discal series of narrow less defined spots; posterior margin broadly yellow: hindwing with the veins towards the abdominal margin and cilia orange red; a curved ill-defined series of narrow purplish white discal streaks. Third joint of palpi brown; thorax beneath greyish; middle of abdomen beneath and sides of the bands and legs orange red.

Female similar, but with darker bluish grey hairs without the orange red subcostal streak and black discal patch.

Expanse, 2½ inches.

Habitat, Darjeeling." (*Moore, l. c.*)

Also occurs in Cachar (*Wood-Mason and de Nicéville*); Sikkim (*Elwes*).

In collections Indian Museum and de Nicéville.

This species occurs commonly in the Nilgiris; it is the* *I. helirius* of Mr. Hampson, which latter seems to be a very distinct insect, and apparently has never been recorded from Indian limits; its description is appended:

Papilio helirius, Cramer, Pap. Exot. iii, pl. lx., fig. D. (1779).

" The body and wings of this Plebeian are beneath of the same colour that is seen on the upperside of the figure given here, with the exception that the little yellow spot is larger and white. Palpi and feet orange. Hab. Surinam." (*Cramer, l. c.*)

In Cramer's figure all the upper surface of both wings, the body, head, and thorax are of an uniform brown, with a rather deep nearly black border and pale cilia. There is a light yellowish spot just outside the cell on the forewing, of which there is no trace in *I. jaina*.

12.—ISMENE MAHINTHA, *MOORE*.

Ismene mahintha, Moore, P. Z. S., 1874, p. 575, pl. lxvii, fig. 4.

" Upperside dark glossy olive brown: base of forewing and middle of the hindwing with abdominal margin, densely clothed with long ferruginous hairs: forewing in male with a small yellow discal spot and a black oblique square patch below the cell: female with two yellow obliquely disposed discal spots; cilia of forewing yellowish, of hindwing ochrey red. Underside dark bluish, fawn-colour spots on forewing as above. Body clothed with yellowish ferruginous hairs.

Expanse, 1¾ inch.

Habitat, Burmah." (*Moore, l. c.*)

Also recorded from Cachar (*Wood-Mason and de Nicéville*).

In collections Indian Museum and de Nicéville.

GENUS V.—HASORA.

Hasora, Moore, Lep. Cey., vol. i, p. 159 (1881).

Hasora, Distant, Rhop. Mal., p. 374 (1882-86).

" Differs from *Badamia* in the forewing being short, broad, and triangular; first subcostal emitted at one-third, second and third at equal distances between it and end of the cell, fourth half-way between third and the end; middle median at two-fourths and lower at three-fourths before end of the cell; hindwing very short, lengthened hindward; apex slightly angular, exterior margin slightly convex, lobe somewhat lengthened, abdominal margin long; second subcostal at one-fourth from the

---

* Mr. de Nicéville informs me that Mr. Hampson admits he was mistaken.

base; disco-cellulars very slender, outwardly oblique, upper shortest, lower slightly concave; radial from their angle, very slender; middle median from close to end of the cell, lower at one-half before the end. Thorax stout; abdomen short.

Type, *H. badra*." (*Moore, l. c.*)

13.—HASORA BADRA, *MOORE.*

Goniloba badra, Moore, P. Z. S., 1865, p. 778.

Hasora badra, Moore, Lep. Cey., vol. i, p. 159, pl. 65, figs. 4, 4, a (1881).

" Male and female yellowish brown.

Male, with a suffused blackish subbasal patch; forewing with three conjugated very small yellowish semitransparent spots near the costa, one-fourth from the apex. Cilia pale greyish brown. Underside brown suffused with purple; forewing with a blackish costal patch before the apex, posterior margin yellowish; hindwing with a subbasal and submarginal suffused blackish band, the latter terminating in a black patch on anal lobe; above the patch is a purple-white streak, and within the cell a small bluish white spot. Palpi and body beneath dull yellow. Legs, pale brown. Female above brown suffused with vinaceous yellowish brown basally; forewing with the three small subapical spots (as in male) and three rather large obliquely quadrate spots, two being disposed on the disc, the third above and within the cell. Underside with the spots on forewing as in upperside; hindwing as in male.

Expanse, ♂ 2, ♀ 2¼ inches.

Habitat, Bengal." (*Moore, l. c. in P. Z. S.*)

Also occurs in the Andamans (*Wood-Mason and de Nicéville*); Sikkim (*Elwes; de Nicéville*); Cachar (*Wood-Mason and de Nicéville*); Ceylon (*Moore*).

I have obtained this species rarely at Rangoon, and at Beeling in N. Tenasserim.

In collections Indian Museum and de Nicéville.

14.—HASORA ANURA, *DE NICÉVILLE.*

Hasora anura, de Nicéville, Journal Bombay Nat. Hist. Soc., vol iv, p. 170, pl. B., figs. 5, ♂; 1♀. (1889).

" HABITAT: Sikkim, Khasi Hills.

EXPANSE: ♂ ♀, 2·1 inches.

DESCRIPTION: MALE. UPPERSIDE, *both wings* deep bronzy-brown, the base and disc thickly clothed with long ochreous-brown hairs; *cilia* ochreous-brown. *Forewing* with a minute subapical transparent shining yellow dot. UNDERSIDE, *both wings* dark brown, somewhat glossed with purple. *Forewing* with the inner margin broadly pale, a broad discal dark band free from purple gloss. *Hindwing* with the basal two-thirds much darker than the outer third, the dark portion well-defined, bearing towards the abdominal margin on the dividing edge a small prominent ochreous spot, an ochreous anteciliary line from the anal angle to the first median nervule, the ochreous spot and line obscure in one specimen; a prominent whitish spot in the middle of the disc in one specimen, obscure in the other. FEMALE. UPPERSIDE, *both wings* coloured as in the male. *Forewing* with a quadrate spot at the end of the cell, an elongate one below across the first median interspace, its inner edge straight, its outer edge concave; another smaller narrow spot constricted in the middle across the middle of the second median interspace; three increasing subapical dots—all these spots shining translucent rich ochreous. UNDERSIDE, *forewing* with the spots of the upperside showing through, the inner margin broadly bright ochreous: otherwise as in the male.

Closely allied to the common *Hasora badra*, Moore, from which it differs in both sexes in having no large anal lobe to the hindwing, this lobe being present in *H. badra* and coloured black on the underside, of which black patch there is no trace in *H. anura;* the latter also is a smaller insect; the female differs in having the three large discal yellow spots of the forewing considerably smaller, and of a deeper richer yellow.

Described from two male and four female specimens in Mr. Otto Möller's collection which shew hardly any variation. They have been selected from ninety-three males and forty-five females of *H. badra*, a very common Sikkim species in Mr. Möller's collection. The complete absence of the large anal lobe or tail in *H. anura* makes it distinguishable from *H. badra* at a glance. There is also a specimen of this species from Sikkim in the collection of Mr. G. C. Dudgeon, and a male from Shillong in the collection of the Indian Museum, Calcutta. This latter specimen was submitted for determination to Mr. F. Moore, who pronounced it to be a variety of *H. badra*, but I believe it to be a good species.

I may note that the *Hasora vitta* of Distant is the *H. coulteri* of Wood-Mason and de Nicéville. A specimen from Perak is in the Indian Museum, Calcutta, and differs from the type male specimens from Cachar in possessing two minute semi-transparent yellow dots on the disc of the forewing, and a similar spot in the discoidal cell of the hindwing on the underside, characters of no importance. The true *H. vitta*, Butler, which is from Sarawak, Borneo, may be known from *H. coulteri* by having the basal area of the hindwing on the underside glossed with green (*virescente*); this is not found in *H. coulteri*." (de Nicéville, *l. c.*)

15.—HASORA HADRIA, *DE NICÉVILLE.*

? *Hesperia badra*, Butler, (*nec* Moore), Trans. Linn. Soc. Lond., Zoology, second series, vol. i, p. 554, n. 3 (1877); *Hasora badra*, Distant (*nec* Moore), Rhop. Malay., p. 374, n. i, pl. xxxv, fig. 3, male (1886).

Hasora hadria, de Nicéville, Journal Bombay Nat. Hist. Soc., vol. iv, p. 172 (1889).

"HABITAT: Perak, ? Malacca.

EXPANSE: ♂, 2·1 inches.

DESCRIPTION: MALE. UPPERSIDE, *both wings* as in *H. anura*, mihi. *Forewing* lacking the subapical yellow dot (this however is a trivial character). UNDERSIDE, *both wings* dull brown, not slightly glossed with purple as in *H. anura*, or strongly so as in *H. badra*, Moore. *Hindwing* with a small anal lobe bearing a black patch, in *H. anura* there is no black patch or anal lobe, in *H. badra* both are large. This species is probably variable with regard to the presence or absence of a white or greyish spot in the cell of the hindwing on the underside, and a white or greyish streak above the anal angle, as in the two allied species above named; Mr. Distant describing a "*var.*" of this species as lacking these characters.

I have not figured this species, as Mr. Distant has done so in his "Rhopalocera Malayana." I have described it from a single male from Perak in the collection of the Indian Museum, Calcutta, which Mr. Distant ticketed "*Hasora badra*, Moore (*var*)." I am unable to say whether or not *H. badra* occurs in Malacca, Johore, Java, Celebes, and the Philippines (localities given for *H. badra* by Messrs. Distant and Butler). The true *H. badra*, Moore, occurs in Sikkim, Assam, Calcutta (one female taken by Colonel G. F. L. Marshall, R. E., in his room at midnight in February), Ceylon, Chittagong, Moulmein and the Andaman Isles (a single female)." (*de Nicéville, l. c.*).

16.—HASORA COULTERI, *WOOD-MASON and DE NICÉVILLE.*

Hasora coulteri, Wood-Mason and de Nicéville, J. A. S. B., vol. lv., pt. 2, p. 378. no. 201, pl. xviii, figs. 8, *male*; 8*a*, 8*b*, *female* (1886).

Hasora vitta, Distant, (*nec Butler*), Rhop. Malay., p. 375, n. 2, pl. xxxv, fig. 4, *male* (1886).

"♂. Upperside, both wings dark bronzy brown, paler at the base owing to the presence of a thick clothing of paler olivaceous brown setæ; costal margin of the forewing and the veins and outer margins of both wings darker and faintly glossed with purple; and with the cilia smoky brown. Forewing without spots, but with three ill-defined discal bands composed of modified scales arranged along each side of the submedian nervure, and of the first and second median nervules, and probably concealed by setæ in the living insect. Underside, forewing strongly glossed with purple at the apex, and with a brownish ashy lustrous patch, extending nearly to the outer margin, divided by the submedian nervure, and slightly diffused over the disc. Hindwing crossed by a pearly-white slightly outwardly concave prominent discal band, which extends from the costal to the submedian nervure, where it is slightly recurved, is broadest in the middle of its length, narrowest at its posterior or inner extremity, and reappears close to the abdominal margin as a clump of white scales divided by the internal nervure; the wing suffused with purple beyond the white band, especially on the dark anal blotch, in front of which there extends nearly as far as the third median nervule a distinct whitey-brown anteciliary line.

♀. Upperside, both wings darker and more richly coloured than in the male, being very conspicuously glossed with purple beyond the lighter basal portions. Forewing with three golden-yellow semi-transparent lustrous spots, two discal, situated and shaped much as in *Parata chromus*, Cramer, and one minute and subapical one at the junction of the first and second thirds of the length of the last subcostal cell, sometimes with two spots one above the other, the upper the smaller, and placed in the interspace next above. Underside, forewing marked as above, with the inner margin ashy-white, and with a patch of ashy scales in the internomedian area at the level of the discal spots. Hindwing with the band broader than in the male and extending to the costal margin. In one of the specimens of this sex the upperside has a mottled appearance owing to the purple gloss being imperfectly developed, as in so many *Euplœas*. Head and thorax covered with iridescent olive-green pubescence nearly concolorous with that of the wings; abdominal segments of the body above and below edged posteriorly with yellowish inclining to orange above; eyes encircled with whitey-brown scales; palpi clothed with mixed brown and yellowish scales.

Expanse, 2·1 inches.

Four males and two females Silcuri [Cachar] and neighbouring forests, June 1st to July 25th."

(Wood-Mason and de Nicéville, l. c.)

In collection Indian Museum.

GENUS VI.—BIBASIS.

Bibasis, Moore, Lep. Cey., vol. i, p. 160 (1881).

"Intermediate between *Parata* and *Hasora*. Forewing comparatively longer than in either of those genera; apex more pointed, exterior margin more oblique and posterior margin shorter than in *P. chromus*; first subcostal emitted at nearly one-half before end of the cell, second at one-fourth before the end; middle median at one-fourth before end of the cell. No glandular streak in male; hindwing more oval, less convex externally, and broader posteriorly. Type *B. sena*." *(Moore, l. c.)*

17.—BIBASIS SENA, *MOORE*.

Goniloba sena, Moore, P. Z. S., 1865, p. 778.

Bibasis sena, Moore, Lep. Cey., vol. i, p. 160, pl. 65, figs. 3. 3 a. (1881).

"Male. Upperside dark chocolate brown. Cilia of hindwing carmine-red. Underside maroon brown; forewing. with a large buff-white patch from the middle of posterior margin, bordered above with purple; hindwing with a broad transverse purple white band terminating before the anal angle, the inner border of which is sharply defined, the outer suffusing itself on the disk. Cilia carmine-red. Palpi and thorax in front, beneath and anal tuft dull yellow. Thorax beneath greyish brown.

Expanse, 2 inches.

Habitat, Bengal." *(Moore, l. c. in P. Z. S.)*

The female is described by Mr. Moore in his "Lepidoptera of Ceylon" as not differing from the male.

Also recorded from Cachar (*Wood-Mason and de Nicéville*); Nilgiris (*Hampson*); Sikkim and Assam (*Elwes*).

In collections Indian Museum and de Nicéville.

GENUS VII.—PARATA.

Parata, Moore, Lep. Cey., vol. i, p. 160 (1881).

" Forewing narrower and less triangular than in *Hasora*, the exterior margin shorter, and posterior margin longer; middle median emitted at nearly one-half, and lower at one-sixth before end of the cell. Male with an oblique glandular streak of laxly raised scales below the cell: hindwing less produced hindward; anal lobe short and angular; second subcostal emitted at one-third from the base; disco-cellulars very oblique; antennæ more slender.

Type *P. chromus*." (*Moore, l. c.*)

The two species of this genus are doubtfully distinct. Typically, *P. alexis* is smaller than *P. chromus*, with broader and more prominently marked band on the hindwing; but Mr. Elwes states that in a large series of specimens from Ceylon, Bangalore, Sikkim, Shillong, Barrackpur, and Burmah, he finds too much variation both in size and in the band of the underside to enable him to distinguish them. Mr. de Nicéville, however, is of opinion that they may be distinct, and that *alexis* occurs only in S. India and Ceylon, in which case the specimens recorded by Mr. Butler and Colonel Swinhoe would probably be referable to *chromus*.

18.—PARATA CHROMUS, *CRAMER*.

Papilio chromus, Cramer, Pap. Exot., vol. iii., pl. 284, fig. E. ♂. (1782).

Goniloba chromus, Moore, P. Z. S., 1865, p. 777.

Parata chromus, Moore, Lep. Cey., vol. i., p. 161, pl. 65, figs. 1. a. b. (1881).

" Male and female dark vinaceous brown.

Male with suffused blackish subbasal patch; both wings greyish brown basally. Cilia greyish brown. Head and thorax greenish brown. Abdomen brown. Underside with the apex of forewing suffused with purple blue; hindwing with a narrow transverse discal bluish white band, a blackish patch on anal lobe, exterior to which the cilia have a short white line. Third joint of palpi and legs brown: palpi and thorax beneath dull yellow.

Female paler brown; forewing with two yellowish semi-transparent discal spots, and a very small similar spot before the apex.

Expanse, 2 inches.

Habitat, Bengal." (*Moore, l. c. in P. Z. S.*)

Also occurs in the Andamans and at Bangalore (*Wood-Mason and de Nicéville*); Calcutta (*de Nicéville*); Ceylon (*Moore*); Hyderabad, Sind (*Swinhoe*); Nilgiris (*Hampson*); Orissa (*Taylor*); Sikkim (*Elwes*). And I have obtained it in Rangoon, Ganjam, Madras, and the Nilgiris.

"Andaman females all have only a single small semi-transparent subapical speck between the last two branches of the subcostal nervure of the anterior wing; but those from continental India have sometimes one and sometimes two besides this on the disc of the same wing, which in one from Bangalore in South India, are enlarged into two conspicuous reversed comma-shaped spots." (*Wood-Mason and de Nicéville*, J.A.S.B., 1881, p. 254).

In a long series of this species from Madras, the females have one very minute subapical spot, and two prominent spots on the disc, the lower of which is indented outwardly. Both males and females show very little variation in the prominence of the transverse band on the underside, though this is not the case in specimens from the Nilgiris, where *alexis* and *chromus* seem to run into one another.

In collections Indian Museum and de Nicéville.

19.—PARATA ALEXIS, *FABRICIUS*.

Papilio alexis, Fabricius, Syst. Ent., p. 533 (1775).

Parata alexis, Moore, Lep. Cey., vol. 1., p. 161, pl. 65, fig. 2a; 2b. (1881).

"Upperside dark vinaceous olive brown, base of both wings olive green. Male: forewing with a suffused blackish subbasal patch traversed by a curved oblique black glandular streak. Underside vinous brown; costal border of forewing and basal area of hindwing suffused with purplish-blue: hindwing with a broad well-defined transverse white band broken above large blackish anal lobe spot.

Female. Forewing with two small semi-transparent yellowish discal angular spots. Underside as in male, the forewing also showing the discal spots and a whitish streak along posterior margin.

Expanse, $1\frac{1}{2}$ to $1\frac{3}{4}$ inches.

Smaller than *P. chromus*, with broader and more prominently marked band on underside of hindwing." (*Moore, l. c.*)

Hab. Ceylon (*Wade, Mackwood*).

Also recorded from Mhow, Poona, Bombay, and Karachi (*Swinhoe*); Tret, N.-W. India (*Butler*); Orissa (*Taylor*); Nilgiris (*Hampson*).

This species seems to differ in the number of spots on the forewing, I have only two females both obtained at Berhampore in Ganjam. Of these one has two largish discal spots and one small round subapical spot; the other specimen has the subapical spot and a similar small round one on the disc corresponding to the upper of the two present in the other specimen; it also has the band on the underside of the hindwing obsolete except towards the anal angle.

In collections Indian Museum and de Nicéville.

20.—PARATA MALAYANA, *FELDER.*

Ismene malayana, Felder, Wien. Ent. Mon., vol. iv, p. 401, n. 28 (1860).
Ismene malayana, Felder, Reise Nov., Lep., vol. iii, t. 72, f. 15 (1866).
Choaspes? malayana, Distant, Rhop. Mal., p. 373, pl. xxxv, fig. 2 (1886).

"Alis supra fuscis, subtus anticarum limbo costali, posticarum dimidio basali chalybæis, his striga discali alba. ♂" (*Felder, l. c.*)

"The females have a small semi-transparent yellowish discal speck between the two posterior branches of the median vein, and of course lack the oblique band of short lines of modified scales seen in the males of this as well as of the preceding closely-allied species." (*Wood-Mason and de Nicéville*, J. A. S. B., 1881, p. 254.)

The above refers to Andaman females only, as in the Nicobar females the small semi-transparent yellow discal speck between the two posterior branches of the median vein is wanting according to Messrs. Wood-Mason and de Nicéville.

Recorded from the Andamans and Nicobars.
In collection de Nicéville.

GENUS VIII.—PADUKA.

Paduka, Distant, Rhop. Mal., p. 375 (1886).

"Anterior wings elongate, subtriangular; costal margin oblique, outer margin nearly straight, inner margin very slightly rounded. Costal nervure terminating on costa a little before the end of cell, first subcostal nervule emitted a little beyond middle of cell; second, third, and fourth subcostal nervules about equal distances apart, fifth from near end of cell; discocellular nervules about equal in length, the upper suberect, the lower obliquely directed inwardly; middle median nervule slightly nearer upper than lower. Posterior wings with the costal margin rounded, the outer margin sinuated and somewhat lobately produced at anal angle; subcostal nervules bifurcating about middle of cell; median nervules with their bases moderately close together. Body robust, pilose, the hairs forming several prominent tufts, of which the most noticeable are three in triangular series, above base of abdomen. Palpi broad and coarsely pilose. Legs strongly pilose beneath. Antennæ somewhat long and slender, with a moderately formed club, its apex attenuated and strongly curved or hooked.

Male. Anterior wings above with a large discal patch of silky hairs extending to base along the median nervure, and an elongate patch of long silky hairs on base of inner margin. Posterior wings above with long silky hairs at base and along submedian nervure, and with two prominent discalelongate glandular pouches—or pseudo-scent glands—situated on the second and third median nervules.

Anterior wings beneath with a long tuft of coarse hairs on the submedian nervure." (*Distant, l. c.*)

21.—PADUKA LEBADEA, *HEWITSON.*

Hesperia lebadea, Hewitson, Ex. Butt., vol. iv, pl. iii, figs. 22, 23

Paduka glandulosa, Distant, Rhop. Malay., p. 376, pl. xxxv, fig. 5 ♂ (1886).

Ismene lebadea, var. *andamanica,* Wood-Mason and de Nicéville, J.A.S.B. 1881, p. 254.

" Male. Wings above dark fuliginous brown ; anterior wings with a large discal patch of dark fuscous silky hairs; posterior wings with the fringe pale ochraceous, and with two pale raised discal elongate glandular pouches situated on the second and third median nervules. Wings beneath paler than above; anterior wings with the disc darkest, the inner area palest, and with a long tuft of coarse pale ochraceous hairs on the submedian nervure; posterior wings with a transverse discal pale ochraceous fascia. Body and legs more or less concolorous with wings. Antennæ blackish, their hooked apices ochraceous.

Exp. wings, ♂ 55 millim.

Hab. Malay Peninsula ; Singapore *(coll. Staudinger)*." *(Distant, l. c.)*

I obtained a single male of this species at Beeling, N. Tenasserim, in April.

In collections Indian Museum and de Nicéville.

The following is the description of the local race *andamanica.*

" Male. Wings above dark brown of a slightly greenish tinge, all without spots. Anterior wings bearing a huge and dense pear-shaped sericeous patch of setæ glossed with greyish-greenish and extending nearly from the bottom of the angle formed at the base of the organ by the subcostal and submedian veins about to the level of the end of the fourth fifth of the length of the latter vein, with all the setæ directed backwards and slightly outwards; with the costal margin purplish ; the outer portion beyond the setulose patch bronzy ; and the cilia pale luteous.

Posterior wings purple-glossed, with two subparallel raised discal longitudinal lines of modified scales attached to the apparently thickened bases of the first and second median veinlets, and with the cilia pale orange.

Anterior wings below bronzy-brown with a patch of brilliant amethyst-purple sparsely irrorated with white scales and extending from the end of the cell nearly to the apex of each organ, and with the basal portion of the wing-membrane behind the median vein and its first branch whitey-brown passing to ashy posteriorly, and with a tuft of brown-tipped yellow setæ arranged longitudinally upon and on each side of the basal half of the submedian vein.

Posterior wings below purple-glossed, darkest over the scent-glands, with an interrupted transverse discal band of white scales from near the abdominal margin to the middle of the organs, where it diffuses itself widely over a diffused patch of amethyst-purple.

Female. All the wings above and below paler and duller and glossed with purple, the anterior ones spotted. Anterior wings suffused with purple on the disc, which bears three semi-transparent yellow lustrous spots of the same size, relative proportions, and shape as in C. [*Unkana*] *attina*, Hewitson, with a fourth smaller and elongate yellow opaque spot placed just in front of the submedian vein rather beyond the middle of the organs.

♂. ♀. Eyes blood-red.

Antennæ purplish brown with the club bright luteous below.

Expanse, ♀, 2·65 inches.

Hab. Andamans.

The patch of setæ on the upperside of the anterior wings, the yellow tuft (which probably serves as a scent-fan) on the underside of the same wings, and the lines of modified scales (which probably cover the scent-glands as they seem soiled as if by some exuding fluid) on the upperside of the posterior wings are structures peculiar to the male sex." (*Wood-Mason and de Nicéville, l. c.*)

"*P. lebadea*, var. *andamanica* differs from *P. lebadea*, Hewitson, described from Borneo, in the male being larger, the cilia of the forewing pale yellow instead of white; and in the presence on the underside of the forewing of a prominent oblique powdery patch of violet scales extending from the end of the discoidal cell to the apex of the wing." (*de Nicéville.*)

GENUS IX.—PIRDANA.

Pirdana, Distant, Rhop. Malay., pp. 369, 376 (1886).

"This genus principally differs from the preceding—*Paduka*—in the following characters :—the upper disco-cellular nervule of the anterior wing is longer than the lower; the first and second median nervules of the same wing are emitted moderately close together and remote from the lower median nervule; and there is a complete absence of the glandular patches and pouches as found in *Paduka*.

This genus will also contain the *Hesperia ismene*, Felder, a Celebesian species." (*Distant, l. c.*)

22.—PIRDANA RUDOLPHII, *ELWES* and *DE NICÉVILLE*.

Pirdana rudolphii, Elwes and de Nicéville, J. A. S. B., 1886, p. 438, n. 150, pl. xx, fig. 6 ♂.

"Male. Upperside, both wings rich brown, tinted with vinaceous. Forewing with the cilia concolorous with the rest of the wing. Hindwing with the cilia orange from the anal angle to the first median nervule, broadest in the middle, from the first median nervule to the apex concolorous with the wing. Underside, both wings with the ground-colour as above. Forewing with the inner margin broadly to the first median nervule pale ochreous, the costa and upper half of the cell and all the veins except the median and submedian nervures and the first median nervule streaked with bronzy-green. Hindwing with the cell and all the veins and abdominal margin streaked with bronzy-green, the cilia at the anal angle orange as above, but slightly broader. Head and body above concolorous with the wings on the upperside, palpi, thorax and underside of body orange.

Antennæ dark brown throughout except the tip of the club on the underside which is paler. No secondary sexual characters.

Expanse, ♂ 2·00; ♀ 2·25 inches.

Allied to *Hesperia ismene*, Felder, from Celebes, from which it differs in having the cilia only of the hindwing on both sides at the anal angle orange: in *H. ismene* the anal angle of the wing is broadly orange, as are also the last three segments of the abdomen on the upperside. It is also allied even more closely to *Pirdana hyela*, Hewitson (Distant, Rhop. Malay., p. 376, n. 1, pl. xxxv., fig. 6, ♀ (1886), which in the extent of the yellow area at the anal angle is about intermediate between it and *P. ismene*, and agree with it in having the body dark brown, not yellow, above. It is also more distantly allied to *C. benjaminii*, Guérin.

A single male from Tavoy.

In Colonel Lang's collection is a very old specimen from Sikkim, without head or abdomen, which we believe to be the female of this species. On the upperside of both wings it is obscurely glossed with green on the basal two-thirds, the forewing is rather broader, the outer margin slightly convex, which in the male is slightly concave; and on the hindwing the orange colouring at the anal angle is rather broader (not nearly so broad as in *P. ismene* or *P. hyela*), and the cilia are throughout orange. Underside paler than in the male, the ground-colour obscure green rather than brown, the orange coloration at anal angle of hindwing as on upperside, but rather broader." (*Elwes and de Nicéville, l. c.*)

In collection Indian Museum. The Rev. Walter A. Hamilton has also obtained a single specimen in the Khasi Hills.

GENUS X.—MATAPA.

Matapa, Moore, Lep. Cey., vol. 1, p. 163 (1881).
Matapa, Distant, Rhop. Mal., p. 379 (1886).

" Forewing elongated, triangular; cell extending two-thirds the wing, broadest across the middle, very narrow at each end; subcostals at equal distances apart; first branch emitted at one-third before end of the cell; disco-cellulars extremely oblique, upper bent inward to subcostal, lower very slender, straight; upper radial from the angle, lower from their middle; upper median emitted from end of cell opposite third subcostal, middle median at one-fifth and lower at nearly three-fifths before end of the cell; submedian curved in the middle. Male with an oblique discal slender linear glandular streak of raised scales; hindwing short, rather broad; exterior margin convex in middle; costal vein slightly arched in middle, extending to apex; second subcostal emitted at one-third before end of the cell; the cell long, broad across the middle; disco-cellulars long and very oblique; no radial perceptible; middle median close to end of the cell, lower at nearly one-fifth before the end; submedian and internal straight. Body moderately stout; palpi thick, flat in front, terminal joint very short, thick, conical and imbedded among the scales; antennæ rather long, slender, club thickish, abruptly bent near end and pointed at tip.

Type, *M. aria.*" (*Moore, l. c.*)

Mr. de Nicéville (J. A. S. B., 1883, p. 84) give a very useful key for the discrimination of the five species of this genus.

"*Matapa subfasciata*, is easily recognised by the distinct markings of the underside.

M. aria. Cilia of both wings yellowish-white. Underside ferruginous, in some specimens inclined to ochreous. The long hairs which clothe the body and base of the wings both above and below are hardly perceptibly iridescent greenish. Anal segment of the female furnished with a very close thick tuft of pale yellow hairs. Expanse averaging about 1·6 inches.

M. shalgrama. Cilia of forewing yellowish-white, of hindwing orange-yellow, shading off into yellowish-brown at the apex. Underside varying from dark ferruginous to bright ochreous. Anal segments of the female with a dark brown tuft of hairs, marked with two paler brown streaks on each side. Expanse averaging about 2·1 inches. Other characters as in *M. aria.*

M. sasivarna. Cilia of forewing greyish-white, of hindwing broadly from anal angle to two-thirds of the margin orange-yellow, thence to the angle brown. Underside dull rich brown, in some lights beautifully glossed with iridescent greenish. Anal segment of the female furnished with a fringe (not a very close thick tuft) of long yellow hairs. Long hairs on body and base of wings brilliant (especially in the females) iridescent green. Expanse averaging about 1·8 inches.

M. druna. Cilia as in *M. sasivarna.* Underside dull rich brown glossed with iridescent greenish, but the apex of the forewing perceptibly lighter brown in the males. Long hairs also iridescent green. Anal tuft of female as in *M. sasivarna.* Expanse averaging about 1.95 inches."

23.—MATAPA ARIA, *MOORE.*

Ismene aria, Moore, P. Z. S., 1865, p. 784.

Matapa aria, Moore, Lep. Cey., vol. 1, p. 164, pl. 66, figs. 1, 1*a*, (1881).

Matapa aria, Distant, Rhop. Mal., p. 378, pl. xxxv, fig. 8 (1886).

"Male and female chocolate brown.

Male. Upperside, pale brown; forewing with a short impressed comma-like grey streak obliquely beneath the cell. Cilia yellowish white. Underside bright ferruginous brown. Palpi ferruginous brown.

Female. Upperside dark chocolate brown without the impressed streak; cilia of hindwing pale orange yellow. Underside bright ferruginous brown.

Expanse, ♂ 1⅚, ♀ 2⅛ inches.

Hab. Bengal." (*Moore, P. Z. S. l. c*).

Also recorded from Ceylon (*Hutchison, Wade, Mackwood*); Andamans (*Wood-Mason and de Nicéville*); Calcutta (*de Nicéville*); Cachar (*Wood-Mason and de Nicéville*); Orissa (*Taylor*); Nilgiris (*Hampson*); Sikkim (*Elwes*).

I obtained this species commonly at Rangoon. The eyes are bright red.

Mr. Elwes states that this species is a much smaller insect than *M. shalgrama*.

It is usually common wherever it occurs. In collections Indian Museum and de Nicéville.

24.—MATAPA DRUNA, *MOORE.*

Ismene druna, Moore, P. Z. S., 1865, p. 784, ♀.

Ismene druna, Wood-Mason and de Nicéville, J. A. S. B., 1881, p. 255 ♂.

"Male. Upperside dark olive brown; forewing with a well defined obliquely curved discal impressed grey streak. Cilia of forewing greyish white, of hindwing orange-yellow. Head palpi and legs beneath ferruginous brown. Underside dark purplish brown."

Expanse, 1⅞ inch.

Hab. Bengal." (*Moore, l. c.*)

Also recorded from the Andamans (*Wood-Mason and de Nicéville*); Sikkim (*de Nicéville; Elwes*); Taoo, 3,000 to 3,500 feet, U. Tenasserim (*Limborg*).

"♀. Differs from the male only in the absence of the sexual streak in the anterior wings. In both sexes of this species the anterior wings are tipped with paler on both sides." (*Wood-Mason* and *de Nicéville, l. c.*)

I have several specimens from Rangoon.

In collections Indian Museum and de Nicéville.

25.—MATAPA SASIVARNA, *MOORE.*

Ismene sasivarna, Moore, P. Z. S., 1865, p. 784.

"Male and female. Upperside dark vinaceous brown.

Male. Forewing with a short impressed comma-like greyish white streak obliquely beneath the cell. Cilia of forewing greyish white, of hindwing broadly, from anal angle to two-thirds of the margin, orange-yellow, thence to the angle brown. Underside dark fuliginous brown; cilia as above. Palpi and body blackish brown; abdomen with slight orange-yellow tuft.

Female as in male but without greyish white streak.

Expanse ♂ 1¾, ♀ 2 inches.

Hab. Bengal." (*Moore, l. c.*)

Also recorded from Sikkim (*Elwes and de Nicéville*).

According to Mr. Elwes this is a rarer insect that *M. shalgrama*, and the female differs from the female of that species in having somewhat longer wings and no sexual streak.

In collections Indian Museum and de Nicéville.

26.—MATAPA SHALGRAMA, *DE NICÉVILLE.*

Hesperia aria, Hewitson (*nec* Moore), Ex. Butt., vol. iv, Hesperia pl. iii, figs. 24, 25, (1868), female.

Matapa shalgrama, de Nicéville, J. A. S. B., 1883, p. 85, n. 28.

"Male. Upperside dull rich chocolate-brown, slightly paler on the outer margin of the forewing. Cilia of forewing yellowish-white, of hindwing orange-yellow shading off into yellowish-brown at the apex. Underside dark ferruginous.

Female. Upperside paler than in the male, the forewing uniformly coloured and lacking the male sexual streak; with the area before the subcostal nervure from the base to half the length of the wing ochreous. Underside lighter coloured than in the male, in some specimens bright ochreous, except the inner margin which is brown extending widely into the disc of the forewing. Anal segment furnished with a very close thick tuft of dark brown hairs, marked on each side with two pale brown bars. Body on the upperside dark brown, below ferruginous or ochreous. Eyes scarlet. Three males and seven females of this species seen by me show but little variation. Hewitson's figure of the female is sufficiently characteristic and make the species easily recognisable.

Expanse ♂ 2·1, ♀ 2·2 inches.

Habitat, Sikkim." (*de Nicéville, l. c.*)

Also recorded by Mr. Elwes from Sikkim.

In collections Indian Museum and de Nicéville, and there is a single specimen in the Phayre Museum, Rangoon, from the Karen Hills, Burma.

27.—MATAPA SUBFASCIATA, *MOORE.*

Ismene subfasciata, Moore, P. Z. S., 1878, p. 686.

Matapa subfasciata, Moore, Lep. Cey., vol. 1, p. 164, pl. 64, fig. 3. a. b. (1881).

"Male. Upperside dark velvety umber brown; costal edge of forewing slightly ochreous; cilia of both wings ochreous. Underside paler: forewing with a pale pink triangular costal patch before the apex; posterior border ochreous, adorned with a large hairy tuft. Hindwing with a transverse pink fascia across middle of the wing. Eyes red. Legs beneath and anal tuft ochreous. Expanse 2 inches.

Hab. Ceylon.

Allied to *I. aria*, Moore, from which it may be distinguished by the markings of the underside." (*Moore, P. Z. S., l. c.*)

"Larva pale purplish-grey, with indistinct darker transverse dorsal lines; head black spotted. Feeds on Palmaceæ. Pupa pale olivaceous-yellow." (*Moore, Lep. Cey., l. c.*)

· Not in collection Indian Museum, and Mr. de Nicéville informs me he has never seen it.

GENUS XI.—CAPILA.

Capila, Moore, P. Z. S., 1865, p. 785.

"Palpi large, porrect, projecting beyond the head, densely pilose; third joint conical half the length of the second. Antennæ extending to half the length of forewing. Body moderately stout. Abdomen extending to near anal angle. Legs slender; femora slightly pilose beneath; hind tibiæ with a dense tuft of very long hairs at the side; mid tibiæ with a pair, and hind tibiæ with two pairs, of apical spurs.

Wings large, broad. Male. Costa, nearly straight; apex acute; exterior margin very oblique; posterior margin abbreviated, half the length of the costa. Hindwing with the apex angled; exterior margin convex, with slight angle in the middle. Female larger. Costa slightly arched; exterior margin oblique; posterior margin two-thirds the length of the costa. Hindwing nearly quadrate, the exterior margin being produced to an abrupt angle in the middle." *(Moore, l. c.)*

28.—CAPILA JAYADEVA, *MOORE*.

Ismene jayadeva, Moore, Cat. Lep. E. I. C., vol. 1, p. 248.
Capila jayadeva, Moore, P. Z. S., 1865, p. 785, pl. xlii, fig. 3.

"Male and female brown. Upperside—base of wings clothed with orange-yellow hairs; both wings with a narrow longitudinal semitransparent streak between the veins, the discoidal cell having two streaks, and a third but short streak arising from the extremity. Thorax, head, and palpi, orange-yellow. Abdomen brown, with narrow white segmental bands; third joint of palpi and a few surrounding hairs and a spot on forehead brown. Underside paler brown, the semitransparent streaks being less prominent. Body and legs brown.

Female similar, but with the thorax and base of wings brown.

Expanse ♂ 2⅝, ♀ 3 inches.

Hab. Darjeeling." *(Moore, P. Z. S., l. c.)*

Also recorded from Sikkim by Mr. Elwes who notes that the female is without the orange on the thorax and base of wings, and has much broader, rounder wings than the male.

One female recorded from Margherita, Assam, by Mr. Doherty.

With reference to this species Mr. A. V. Knyvett writes as follows:—

"I flushed *Capila jayadeva* ♀ off the underside of a leaf in a damp shady spot full of undergrowth. She flitted about like a *Plesioneura* for some time, and then settled on the underside of a broad leaf, with wings outspread. It was an impossible sort of a place to use a net on and I missed, with the result that she flew a short way and again settled in the same way and gave me as easy a chance of taking her as I could have wished for. The flight seemed a compromise between that of a *Mycalesis* or *Yphthima* and a *Plesioneura*, rather incliming towards the latter."

In collections Indian Museum and de Nicéville.

GENUS XII.—PISOLA.

Pisola, Moore, P. Z. S., 1865, p. 735.

" Palpi large, erect, projecting beyond the head, densely pilose; third joint minute, conical. Antennæ rather long, curved backward at the apex. Body very stout; abdomen extending to within one-third of the length of hindwing. Legs moderately slender; femora pilose beneath: mid tibiæ armed with a pair and hind tibiæ with two pairs of slender apical spurs. Wings large, broad; costa of forewing slightly arched: exterior margin oblique; posterior margin straight. Hindwing convex at the base of anterior margin; apex, exterior margin, and anal angle convex. Subcostal vein of forewing six branched; second and third arising at equal distances from the first; fourth to sixth contiguous at their base to the third." *(Moore, l. c.)*

Mr. de Nicéville informs me that the subcostal vein in this genus has only four and not six branches.

29.—PISOLA ZENNARA, *MOORE.*

Pisola zennara, Moore, P. Z. S., 1865, p. 786, pl. xlii, f. 4.

" Male and female. Upperside brown; forewing with a broad yellowish-white semitransparent irregular margined discal band obliquely from middle of costa to posterior angle; hindwing in the male exteriorly with two greyish longitudinal streaks between each vein, these being absent in the female. Abdomen with pale greyish anal tuft. Underside uniform brown, with oblique discal band as above. Front of head and palpi dull orange yellow. Body and legs brown. Cilia brown.

Expanse ♂ 2¼, ♀ 3¾ inches.

Hab. N.-E. Bengal." *(Moore, l. c.)* April to October. Trop.

Mr. Elwes records this species as occurring rarely in Sikkim from April to August in the low valleys, and also states that the antennæ of the female both in this species and in *C. jayadeva* are much less hooked at the tip than those in the male.

In collections Indian Museum and de Nicéville.

GENUS XIII.—PITHAURIA.

Pithauria, Moore, P. Z. S., 1878, p. 689.

Pithauria, Distant, Rhop, Mal., p. 378. (1886).

" Forewing elongated, narrow; apex pointed; exterior margin **very** oblique; hind margin short; hindwing convex externally, lobular at **anal** angle. Head and thorax very broad, robust; abdomen not so long as hindwing. Antennæ with a slender club and very long whip-like tip. Venation similar to *Pamphila*." *(Moore, l. c.)*

30.—PITHAURIA MURDAVA, *MOORE*.

Ismene murdava, Moore, P.Z. S., 1865, p. 784.
Pithauria murdava, Distant, Rhop. Mal., p. 378, pl. xxxv, fig. 9, ♂ (1886).

"Upperside olive brown: forewing with the base grey, with six small yellow spots, two within the extremity of the cell, two near the costa, one-third from the apex, and two midway beneath; hindwing grey to beyond the middle. Underside pale yellowish-brown; disk of forewing blackish, spots as above: hindwing with indistinct submarginal and discal pale yellowish spots. Abdomen above with greyish brown segmental bands. Palpi, abdomen, and legs beneath dull yellow.

Expanse, 2 inches.
Hab. Darjeeling." (*Moore, l. c.*)
Also recorded from Sikkim by Messrs. de Nicéville and Elwes.
I have obtained this species at Beeling, U. Tenasserim.
In collections Indian Museum and de Nicéville.

31.—PITHAURIA STRAMINEIPENNIS, *WOOD-MASON and DE NICÉVILLE*.

Pithauria stramineipennis, Wood-Mason and de Nicéville, J. A. S. B., 1886, p. 388, pl. xv, fig. 5 ♂.
Pithauria murdava, Distant, Rhop. Malay., p. 371, ♀ only (1886).

"♂. Upperside, both wings marked precisely as in *P. murdava*, Moore, but all the setæ on the base of the wings clear whitey-brown with a touch of yellow on all those in front of the submedian nervure of the forewing, those on the interno-median area of this wing being concolorous with the whitey-brown down of the hindwing, the costal area of which is above more or less extensively pale brown. In *P. murdava*, the setæ in the hindwing are yellowish-olivaceous, all those of the forewing distinctly yellower; and the costal area of the hindwing is dark. All the spots and streaks of both sides are no less variable in *P. stramineipennis* than they are in *P. murdava*, so we have not attempted to describe them.

♀. Differs from male in being larger, in the wings being paler, with the scanty setulose clothing at their bases greyish-fuscous paler than the ground in the hindwing, and in the spots of the forewing being larger, paler, and more angular; agrees therewith in the costal area of the hindwing being pale brown above.

Expanse, ♂ 1·8 to 2·0; ♀ 2·1 inches.
Habitat: ♂. Sikkim, Bhutan, Upper Assam, Cachar; ♀ Sikkim."
(*Wood-Mason and de Nicéville, l. c.*)

In describing this species as above Messrs. Wood-Mason and de Nicéville append the following remarks: "In our figure the downy clothing of the upperside of the wings at the base is not represented of a sufficiently light and bright shade; it is in reality of a clear bright whitey-brown or straw-colour, which, being conspicuously contrasted with the dark margins, renders *P. stramineipennis* most readily distinguishable from *P. murdava*, in which the downy clothing is, as has already been stated, yellowish-olivaceous.

The genital armature, which has been carefully examined in several specimens of each species, though identical in general plan, yet differs greatly in detail in the two.

Several hundreds of specimens of each species have passed through our hands."

Also recorded from Ponsekai; Tavoy; (*Elwes and de Nicéville.*)

In collections Indian Museum and de Nicéville.

GENUS XIV.—PITHAURIOPSIS.

Pithauriopsis, Wood-Mason and de Nicéville, J. A. S. B., 1886, p. 387.

" Male.—Closely allied to *Pithauria*, Moore, but differing, in the forewing, in the distance between the origins of the second and third median nervules being greater, instead of less than that between those of the first and second, in the submedian nervure being strongly sinuated when it comes into relation with, and the internal area expanded opposite to, a prominent bilobed discal glandular organ, extending from the root of the first median nervule for a short distance into the internal area, and consisting of two unequal slight depressions of the wing-membrane, separated from one another by the interno-median fold, and converted by over-arching stiff modified scales into pouches, which are filled with a soft, fine, adhesive brown woolly substance ; and, in the hindwing, in the first subcostal nervule being at its origin strongly arched towards the costal, and the base of the second slightly bowed into the cell and more acutely angled at its junction with the disco-cellular nervule, in the subcostal, in fact, with its branches having the shape rather of a tuning-fork than of the letter Y ; discoidal nervule absent, and only one disco-cellular consequently present, as in *Pithauria*.

The male genital somites and appendages, though at first sight appearing very different, yet when carefully examined are seen to be built on the same plan and to differ in characters of specific value only from those of *Pithauria*.

Female unknown." (*Wood-Mason and de Nicéville, l. c.*)

32.—PITHAURIOPSIS AITCHISONI, *WOOD-MASON and DE NICÉVILLE.*

Pithauriopsis aitchisoni, Wood-Mason and de Nicéville, J. A. S. B., 1886, p. 387, pl. xv, fig. 4 ♂.

" ♂. Upperside, both wings rich bronzy-brown. Forewing with a basal streak on the costa, another in the cell, one in the submedian, and a fourth, the longest of all, filling the interno-median interspace, all composed of long yellowish-olivaceous hair-like scales; two oblong spots placed obliquely at the end of the cell, the anterior one the further from the base of the wing; two subcostal spots, the anterior one a mere dot; an oval or heart-shaped spot in the second median interspace; and a rhomboidal one in the space behind, all semi-transparent ochreous. Hindwing with all but the costal and outer margins clothed with very long yellowish olivaceous hairs. Underside, forewing blackish, with the costal and apical half of the wing yellowish-brown. The spots as above save that those in the cell are conjoined into an ill-shaped figure of 8, and that there is a submarginal series of yellowish dots from the first median nervule to the costa. Hindwing with a minute yellow spot in the subcostal interspace, two large ones (the anterior the larger) in the median interspaces, a submarginal series of small obscure yellowish spots, the two of which in the submedian interspace are large and prominent, and a yellow streak on the abdominal margin. Cilia cinereous on the hindwing and at the anal angle of the forewing, in front of which point they are dark brown.

Expanse, 1·8 inches.

Two males, Irangmara, [Cachar] 6th and 29th July.

This species agrees in markings almost exactly with the male of *Pithauria murdava*, Moore, but the large and curiously-formed sexual brand in the male will at once distinguish it." (*Wood-Mason and de Nicéville, l. c.*)

In collection Indian Museum; and in the Phayre Museum, Rangoon, there are numerous specimens from the Karen Hills.

GENUS XV.—BAORIS.

Baoris, Moore, Lep. Cey., vol. i., p. 165 (1881).

Baoris, Distant, Rhop. Mal., p. 379 (1886), part.

"Forewing triangular; apex acute; exterior margin very oblique: hindwing broad, very convex exteriorly, the male possessing a more or less prominent tuft of long hair covering a patch of raised scales at end of the cell. Body robust, thorax very broad; club of antennæ somewhat lengthened.

Type *B. oceia* (*Hesperia oceia*, Hewitson)." (*Moore, l. c.*)

33.—BAORIS OCEIA. *HEWITSON.*

Hesperia oceia, Hewitson, Desc. Lep. Hesp., 1868, p. 31.

Hesperia cahira, Moore, P. Z. S., 1877, p. 593, pl. lvii, fig. 8 (♀ only.)

Hesperia oceia, Wood-Mason and de Nicéville, J. A. S. B., 1881, p. 258.

Baoris oceia, de Nicéville, J. A. S. B., 1883, p. 85, pl. x, fig. 11 ♀.

Baoris unicolor, Moore, P. Z. S., 1883, p. 533.

Baoris scopulifera, Moore, P. Z. S., 1883, p. 332.

"Male. Wings above rich dark purple-brown with bronzy reflections. Anterior wings typically with eight semi-transparent pale yellow lustrous spots, namely, two, dot-like, at the end of the cell, of which the posterior is the larger, a third subquadrate, the largest of all, between the first and second median veinlets, a fourth, about half the size, between the second and third median veinlets, with a dot, the fifth, beyond and in front of it, and a series of three dots, the sixth, seventh and eighth, in a series, in front of this again.

Posterior wings each with a conspicuous tuft of long dark brown pale-based setæ inserted into the wing membrane immediately behind the base of the subcostal trunk.

Wings below lighter and duller.

Anterior wings with a huge oval ashy patch of a most brilliant satiny lustre, occupying the middle four-fifths of the portion of the organs between the median vein and the posterior margin, and in middle of which is so placed as to be divided by the submedian vein a very much smaller oval patch of brown modified scales. Female, wings above paler and scarcely at all suffused with purple, with the setæ olive-green and the cilia pale luteous. Anterior pair all but invariably with nine spots, an additional opaque one being present just in front of the submedian vein a little beyond the middle of the organs. Wings below pure dead uniform olive-brown.

Expanse, ♂ 1·63, ♀ 1·88 inches.

The female has been described by Mr. Moore as that of his *H. cahira*." (*W.-M.* and *de N. l. c.*).

According to Messrs. Wood-Mason and de Nicéville the male in Andaman specimens varies considerably in the number of spots while the female is almost constant. In the male, if the semicircular series of eight discal spots, be numbered in the order of their succession from before backwards, *inwards* and forwards, then while the third, fifth, sixth and seventh spots are invariably present, the first and the fourth are frequently absent, and the second and eighth occasionally so.

Hab. Andamans, Sikkim.

In recording this species from Sikkim (J. A. S. B., 1883, p. 85) Mr. de Nicéville notes that specimens from that locality vary even more than those from the Andamans, and range "from totally unmarked specimens of both sexes through every gradation to the typical number of eight spots." Also recorded from Cachar (*Wood-Mason and de Nicéville*); Tavoy (*Elwes and de Nicéville*); Calcutta (*de Nicéville*); Orissa (*Taylor*); Sikkim (*Elwes*).

Two varieties of this species have been described by Mr. Moore as *B. unicolor* and *B. scopulifera*, of which the former is the spotless form which is figured by de Nicéville as quoted above, and the other has six spots in the male and eight in the female.

In collections Indian Museum and de Nicéville.

34.—BAORIS PENICILLATA, *MOORE*.

Baoris penicillata, Moore, Lep. Cey., vol. i, p. 166, (1881).

" Allied to *B. scopulifera*. Upperside dark olive brown. Male: forewing with four semi-diaphanous yellow spots, two being apical and two discal, which are also smaller; hindwing with a black tuft. Underside of forewing marked as on upperside, and with a glossy purple space on hind margin enclosing a small brown patch of raised scales.

Expanse, 1⅜ inch." (*Moore, l. c.*)

Recorded from Ceylon (*Mackwood.*)

Not in collection Indian Museum, unknown to Mr. de Nicéville.

GENUS XVI.—CHAPRA.

Chapra, Moore, Lep. Cey., vol. i, p. 168 (1881).

Baoris, Distant, Rhop. Mal., p. 379 (1886), part.

"From typical *Gegenes* (*G. nostrodamus*) this genus differs in its somewhat lengthened form of forewing, more lobular anal angle of hindwing, longer antennae—which have a whip-like joint, and the male insect in possessing an oblique glandular streak below the cell, somewhat as in *Pamphila comma*.

Type, *C. mathias.*" (*Moore, l. c.*)

35.—CHAPRA MATHIAS, *FABRICIUS*

Hesperia mathias, Fabricius, Ent. Syst., Supplt., p. 433 (1798).

Hesperia julianus, Latreille, Enc. Meth., vol. ix, p. 763 (1823), ♂.

Gegenes thrax, Hübner, Samml. Exot. Schmett. vol. ii, pl. 150, f. 1-4 (1820-26).

Chapra mathias, Moore, Lep. Cey., vol. i, p. 169, pl. 70, figs. 1a. (1881).

Baoris mathias, Distant, Rhop. Mal., p. 380, pl. xxxv, fig. 10, (1886).

Chapra mathias, de Nicéville, Journ. Bomb. Nat. Hist. Soc., vol. iv, p. 176, n. 14, pl. B, fig. 7, *male*, (1889).

"Male. Upperside olive brown : forewing with two small yellowish semi-transparent spots within end of cell, three before the apex, and in the male three oblique discal spots followed by a dark-bordered slender straight impressed glandular streak : hindwing with one or two very indistinct pale discal spots. Female with five discal spots in the forewing, and four or five in the hindwing. Underside paler ; markings more distinct ; hindwing also with a spot at upper end of the cell. Expanse 1⅜ to 1⅝ inch." (*Moore, l.c.*)

Recorded from Ceylon ; Sikkim ; Andamans ; Kumaon ; Cachar ; Upper Burmah ; Moulmein to Meetan, Upper Tenasserim ; Mhow ; Bombay ; Poona ; Ahmednagar ; Karachi ; Kangra, N. W. Himalayas, Orissa, Nilgiris.

Occurs commonly throughout India.

In collections Indian Museum and de Nicéville.

Mr. Elwes states this species can be distinguished in the male sex by the brand on the forewing characteristic of the genus, and in both sexes by a spot on the middle of the underside of the hindwing near the base ;

in some specimens one or two spots showing on the upperside of the hindwing. He also adds that he is unable to distinguish it from *C. agna*, of which he has numerous specimens from the Bombay Presidency.

36.—CHAPRA SUBOCHRACEA, *MOORE.*

Pamphila subochracea, Moore, P. Z. S., 1878, p. 691.

"Upperside glossy luteous olive-brown; cilia yellowish-cinereous.

Male. Forewing with two pale semi-diaphanous spots at end of the cell, three contiguous spots obliquely before the apex, three upper discal spots, below which is a narrow white oblique streak or brand; hindwing with three small yellow upper discal spots, the two lowest small.

Female. Forewing with a lower or fourth discal spot, and a small dot below the third spot; the spots angled outward; hindwing as in male. Underside greenish-ochreous, brown on hind border of forewing and anal lobe; marginal line brown and prominent: forewing with the lower spot diffused and white; hindwing with the upper discal white spot large and quadrate, four spots below in a slightly linear position, the upper spot indistinct; a white spot also at upper end of cell, and a smaller indistinct spot above it.

Expanse, ♂ 1⅜, ♀ 1¾ inch.

Hab. Calcutta.

Allied to *P. mathias*, Fabricius, but of large size and more prominently marked." (*Moore, l. c.*)

Also recorded from Cachar (*Wood-Mason and de Nicéville*); Calcutta (*de Nicéville.*)

This species is doubtfully distinct from *C. mathias*.

In collection Indian Museum.

37.—CHAPRA AGNA, *MOORE.*

Hesperia agna, Moore, P. Z. S., 1865, p. 791, ♂

Hesperia chaya, Moore, P. Z. S., 1865, p. 791, ♀

Chapra agna, Moore, Lep. Cey., vol. i, p. 169 (1881).

Baoris chaya, Distant, Rhop. Mal., p. 380, pl. xxxv, fig. 9 (1886).

"Upperside glossy olive-brown; forewing with a series of six very small rather indistinct whitish semi-transparent spots curving from before the apex to the middle of the wing; beneath these is a short oblique pale impressed streak, which is suffused with black on its anterior margin. Cilia pale brown. Underside pale brown; spots on forewing as above but less defined; hindwing with a curved discal series of white dots and a single dot near the base. Palpi and body beneath pale brownish-yellow.

Expanse, 1¼ inch.

Habitat, Bengal." (*Moore, l. c. in P. Z. S.*)

"Similar to *C. mathias*; differs in being somewhat larger, much darker and more uniformly coloured; the spots on the forewing of the male much smaller, those on the disc being slender (not quadrate as in C. *mathias*), and the oblique glandular patch not so prominent. Female also similar." (*Moore*, l. c. in Lep. Cey.)

Recorded from Ceylon (*Hutchison*); Poona, Mhow, Belgaum, Bombay (*Swinhoe*); Nicobars (*Wood-Mason and de Nicéville*); Calcutta (*de Nicéville*); Cachar (*Wood-Mason and de Nicéville*); Nilgiris (*Hampson*.)

This species is also doubtfully distinct from C. *mathias*.

In collection Indian Museum.

38.—CHAPRA PROMINENS, *MOORE*.

Chapra prominens, Moore, P. Z. S., 1882, p. 261.

"Male and female. Upperside dark olive brown, basal area brighter olive. Male—forewing with eight rather large quadrate yellowish semi-diaphanous spots, three being disposed before the apex, three discal and two very obliquely at end of the cell; a prominent narrow oblique yellow brand or streak below the cell, which in the female is replaced by two spots, the upper one of which is very small: hindwing with four yellow semi-diaphanous contiguous spots. Underside paler; spots on forewing as above, the brand showing as a diffused yellow patch from its outer edge; the series of spots on hindwing more prominently white with a fifth spot at the upper end, and one also at the upper end of the cell.

Expanse, $1\frac{3}{5}$ inch.

Hab. N.-W. Himalayas: Tonse valley 6,000 ft., Gurwhal (*Lang*); Kussowlee; Kangra.

An allied species to this from Shanghai, N. China, has lately been described by Mons. Mabille as *Gegenes sinensis* (Bull. Soc. Zool. de France, 1887, p. 232) from which the above differs in its somewhat broader wings and larger size of the markings on the forewing." (*Moore, l. c.*)

Also recorded from Sikkim (*de Nicéville: Elwes*); Kumaon (*Doherty*); Nilgiris (*Hampson*).

In collections Indian Museum and de Nicéville.

39.—CHAPRA NOSTRODAMUS, *FABRICIUS*.

Papilio nostrodamus, Fabricius, Ent. Syst., vol. iii., pt. 1, p. 323, n. 246 (1793).

Pamphila nostrodamus, Kirby, Eur. Butt., p. 65 (1882).

"Dark brown, base smoky black; inner margin of the hindwing paler than the ground colour, and a few white dots on the forewing in the female. Underside pale brown, with some obscure white spots towards the tip of the forewing, and in the female at the hind-margin of the hindwing also. Expands a little over one inch. It inhabits South Europe, North Africa, and Western Asia in August, and is found in dry places." (*Kirby, l. c.*)

Recorded from Campbellpore, Kala Pani and Hurripur, N.-W. India, (*Butler*).

In collection Indian Museum.

40.—CHAPRA KARSANA, *MOORE.*

Hesperia karsana, Moore, P. Z. S., 1874, p. 576, pl. lxvii, fig. 6.

"Upperside pale olive brown. Cilia pale fawn-colour. Male—forewing with minute oblique subapical spots more distinct below, where is a discal row of four somewhat quadrate spots, the third spot smallest.

Underside much paler, marked as above; space from abdominal margin to middle of wing pale brownish white.

Expanse, 1⅛ inch.

Habitat, Rawul Pindi, N. Punjab. (*Capt. H. B. Hellard*)." (*Moore, l. c.*)

Recorded from Kumaon (*Doherty*); Campbellpore, N.-W. India (*Butler*); Quetta, Kandahar and Karachi (*Swinhoe*).

According to Col. Swinhoe this is "a common species at Karachi at all seasons of the year."

In collections Indian Museum and de Nicéville.

A slight variety of this species has been described as below.

Hesperia karsana, var. *saturata*, Wood-Mason and de Nicéville, J. A. S. B., 1882, p. 19.

"Much darker and without a trace of spots on the upperside. One female from Kamorta [Nicobars]; and Kulu, N.-W. Himalayas." (*Wood-Mason and de Nicéville, l. c.*)

I obtained a single specimen of this variety at Quetta, it is unmarked on upperside, but the usual spots are distinctly traceable on the underside of the forewing.

In collection Indian Museum.

GENUS XVII, PARNARA.

Parnara, Moore, Lep. Cey., vol. 1, p. 166 (1881).
Baoris, Distant, Rhop. Mal., p. 379 (1886), part.

"This genus comprises a group of species which have been variously referred to *Hesperia*, *Pamphila* and *Gegenes*. They are similar in form of wings, venation, and antennæ to *Chapra*, but the males have no oblique glandular streak on the forewing.

Type, *P. guttata*." (*Moore, l. c.*)

41.—PARNARA GUTTATA, *BREMER and GREY.*

Eudamus guttatus, Bremer and Grey, Schmett. N. China, p. 10 (1853).
Pamphila mangala, Moore, P. Z. S., 1865, p. 792.
Hesperia bada, Moore, P. Z. S., 1878, p. 688.
Parnara bada, Moore, Lep. Cey., vol. 1, p. 167, pl. 70, figs. 2, 2a (1881).

"Upperside dark brown; forewing with six small whitish semi-transparent spots curving before the apex, also two small similar spots within the cell; hindwing with a discal linear series of four conjugated semi-transparent spots. Cilia pale brownish white. Underside paler, suffused with greenish yellow; forewing with spots as above: hindwing with an additional spot (which is not semi-transparent) at the extremity of the cell. Palpi and body beneath dark yellowish green.

Expanse, 1¼ inch.

Hab. Bengal." (*Moore. l. c.* in P. Z. S., 1865).

Recorded from Murree, Thundiani and Hassan Abdul, N.-W. India (*Butler*); from Kangra District, N.-W. Himalayas (*Moore*); and from Sikkim (*Elwes*).

Mr. Elwes treats this species as synonymous with *P. bada*. He states he has specimens from Shanghai, Japan, Kashmir, Mandi and Sikkim which agree very well. They may be known by the nearly straight line of four transparent spots on the hindwing, and the curved discal series of six or sometimes only five spots on the forewing, as well as two, one or both of which are sometimes wanting, in the cell. Typical *P. bada* (of which he has specimens from Cachar, Bombay, Poona, Sikkim and Ceylon) is distinguished by the less conspicuous markings on the hindwing, usually smaller size, and absence of the two spots in the cell. But he is not sure that these characters are constant.

According to Mr. Butler (P. Z. S., 1886, p. 377) *P. mangala* has the lowest spot on the primaries, larger and more quadrate than in *bada* or *bevani*, and also has the row of spots on the secondaries larger and more nearly in line.

In collections Indian Museum and de Nicéville.

The description of *P. bada* is appended below:

"Allied to *H. mangala*.

Male and female. Upperside dark brown, base of wings olive brown; cilia pale cinereous brown: forewing with two (in some specimens three) contiguous subapical small white semi-diaphanous spots, and three spots below obliquely on the disk, the upper one the smallest: hindwing with a discal irregular linear series of three or four white semi-diaphanous spots, more or less indistinct. Underside greyish brown; both wings marked as above, the spots on the hindwing being more prominent.

Expanse ♂ $1\frac{2}{10}$, ♀ $1\frac{3}{10}$ inch.

Hab. Ceylon." (*Moore*, P. Z. S., 1878, p. 685).

Recorded from Sikkim (*de Nicéville, Elwes*); Calcutta (*de Nicéville*); Kumaon (*Doherty*); Cachar (*Wood-Mason and de Nicéville*); Orissa (*Taylor*); Poona, Belgaum, Bombay, (*Swinhoe*); Nilgiris, (*Hampson*).

42.—PARNARA CINGALA, *MOORE*.

Parnara cingala, Moore, Lep. Cey., vol. 1, p. 167, pl. 70, fig. 3, *a. b.* (1881).

"Allied to *P. bada*. Male and female: forewing with two small semi-diaphanous white spots within end of the cell, three subapical and four discal: hindwing with two very indistinct discal spots. Underside paler;

forewing marked as above, the lowest discal spot prominent and yellow: hindwing with three prominent spots, the upper one being between subcostals.

Expanse $1\frac{3}{10}$ inch.

Larva very pale olivaceous blue, with a darker dorsal and a paler lateral longitudinal line; head yellowish. Feeds on Graminaceæ. Pupa pale olive green.

Hab. Ceylon." (*Moore, l. c.*)

In collection Indian Museum.

43.—PARNARA COLACA, *MOORE.*

Hesperia colaca, Moore, P. Z. S., 1877, p. 594, pl. lviii, fig. 7.
Parnara colaca, Elwes, Trans. Ent. Soc., 1888, p. 446, fig. 1.

"Male and female. Dark olive brown; cilia cinereous; forewing with a recurved discal series of seven small yellow spots, the second from hind margin being the largest; a small spot also at end of the cell. Underside brown, apex and hindwing speckled with olive green scales; a median discal series of small spots on hindwing. Near to *H. cinnara*.

Expanse ♂ $1\frac{3}{10}$, ♀ $1\frac{3}{10}$ inch.

Hab. S. Andamans, Port Blair." (*Moore, l. c.*)

Also recorded from Nicobars and Cachar (*Wood-Mason and de Nicéville*); and Sikkim (*de Nicéville; Elwes*).

According to Mr. Elwes this species is very close to *P. bevani*, from which it may be known by its longer forewing and differently shaped hindwing and by the spots on the upperside of forewing and underside of hindwing being different in number and position. Mr. Elwes gives two woodcuts to illustrate his remarks, and from them it appears that in *P. colaca* the hindwing is more produced at the anal angle that in *P. bevani*, which gives it the appearance of being comparatively narrower, the spots on the upperside of the forewing are almost identical, those in *P. colaca* being slightly larger. On the upperside and underside of the hindwing in *P. colaca* there are only three spots and in *P. bevani* five, but as Mr. Elwes remarks these characters are not constant, still he is inclined to the opinion that the two species are distinct.

In collections Indian Museum and de Nicéville.

44.—PARNARA BEVANI, *MOORE.*

Hesperia bevani, Moore, P. Z. S., 1878, p. 688.
Parnara bevani, Elwes, Trans. Ent. Soc., 1888, p. 447, fig. 2.

"Male. Upperside dark olive-brown; cilia pale brownish-cinereous; forewing with a single small, pale white, semi-diaphanous spot at upper end of the cell, three contiguous subapical spots, another below these; and two larger spots below obliquely on the disc, a small spot also very indistinctly visible on middle of submedian vein; hindwing without spots. Underside greyish olive-brown; spots slightly more prominent than above; hindwing with a discal series of five small somewhat indistinct white spots.

Expanse 1 ⅜ inch.

Hab. Salween, Moulmain (*Lt. Bevan*).

A male specimen of this species from Calcutta collected by the late Mr. Atkinson is in the collection of Dr. O. Staudinger." (*Moore, l. c.*).

Also recorded from Karachi, Mhow, Poona and Bombay (*Swinhoe*); Calcutta (*de Nicéville*); Orissa (*Taylor*); Nilgiris (*Hampson*).

Mr. Elwes also states that he has specimens from Mandi, N.-W. Himalayas, Khasias, and Sikkim.

In collections Indian Museum and de Nicéville.

45.—PARNARA ASSAMENSIS, WOOD-MASON and DE NICÉVILLE.

Parnara assamensis, Wood-Mason and de Nicéville, J. A.S.B., 1882, p. 65.

Parnara assamensis, Wood-Mason and de Nicéville, J. A. S. B., 1886, p. 382, pl. xviii, fig. 5, 5a ♂, pl. xvii, fig. 7, 7a, ♀.

" ♂, ♀. Upperside, both wings rich dark vandyke-brown prominently marked with semi-transparent white lustrous spots, with the base of the interno-median and the basal three-fourths of the inner margin of the forewing, and the posterior or inner half of the hindwing from the base nearly to the outer margin along the veins, clothed with olive-brown setæ. Forewing with ten spots in the male and eleven in the female, *viz.*,—two oblong at the end of the cell, disjunct in the male, but connected at their inner and opposite ends in the female, three subapical, and five discal in the male, but six in the female, forming an oblique series extending from the submedian nervure to the hinder discoidal nervule in the male, but to the subcostal (or front discoidal) in the female; of which spots the first is subtriangular, touches the submedian nervure, and is subequal to the fourth; the second, in the same interspace with the first, is equal to the first subapical, and lies close to, but does not touch, the first median nervule; the third, the largest of all, is equal to, or rather larger than, the first and fourth put together, and acute-angled at its outer end; the fourth is rhomboidal; the fifth rather larger than the second; and the sixth, present in the female only, is shaped somewhat like one of the strokes of a section-sign (§). Hindwing with a small oval discal spot sometimes accompanied by a very minute dot in front of the third median nervure. Underside, forewing marked as above, but with the hindermost spot touching the submedian nervure outwardly diffused. Hindwing covered with minute olive-brown scales, and lighter internal to a straight line drawn obliquely across the interno-median area from the apex of the third median nervule to the base of the submedian nervure, with a prominent white lustrous dot near the anterior end of the cell and a curved discal series of four white lustrous spots, of which the third from the inner margin is by far the largest, transparent, visible on the upper side, and may or may not be accompanied by a minute dot, which may or may not be transparent and visible from above. Body clothed above with olive-green scales and setæ, lighter below. Palpi mixed luteous and olive-green. Antennæ black, broadly half-ringed below before the club with white, with their taper tips ashy.

Expanse, ♂ 2·2 ; ♀ 2·3 to 2·4 inches.

One male in forest near Silcuri [Cachar] on 11th July. It occurs also in Sibsagar (*S. E. Peal*), at Shillong, and in Sikkim, where it is quite a common species at low elevations. It is the largest species of the genus known to us." (*Wood-Mason and de Nicéville*, J. A. S. B., 1886, *l. c.*)

Also recorded from Kumaon (*Doherty*) ; and Sikkim (*Elwes*).

In collections Indian Museum and de Nicéville.

46.—PARNARA ORNATA, *FELDER*.

Hesperia ornata, Felder, Reise Novara, Lep., vol. iii, p. 515, n. 900, pl. lxxii, fig. 6, *male* (1867.)

Parnara ornata, Wood-Mason and de Nicéville, J. A. S. B., 1886, p. 382, pl. xviii, fig. 7, 7a ♂.

" ♂. Wings above shining fuscous, dusted at the base with yellowish grey and hairy. Forewing with a partly divided spot in the cell, also two of median size and three small ones only divided by the veins all subhyaline. On the underside the wings are brown, slightly tinged with violet. The forewing close to the cell is faintly dusky, the hindwing has two spots subovate in front, with a third inside them, and six others silvery white in an oblique series beyond the cell, with a bluish diffused border, a dot at end of cell and some small external spots violet white. Palpi and thorax pale yellow.

Expanse ; 1·9 inches, [from figure].

Habitat ; Buitenzorg, Java." (*Felder, l. c.*)

A single male is recorded from Doarband, Cachar, by Messrs. Wood-Mason and de Nicéville, l. c.

In collection Indian Museum.

47.—PARNARA UMA, *DE NICÉVILLE*.

Parnara uma, de Nicéville, J. A. S. B., 1888, p. 292, pl. xiii, fig. 9 ♀.

" Female. Upperside, both wings rich dark glossy brown, the base clothed with somewhat long greenish-ochreous setæ. Cilia ochreous-brown. Forewing with a spot in the discoidal cell divided in the middle by a fold of the wing, its upper portion lengthened, inwardly sharply pointed ; three increasing conjoined subapical spots, the posterior one nearly twice as large as the other two taken together ; a quadrate spot near the middle of the second median interspace, a larger one in the first median interspace placed exactly midway between the spot in the second median interspace and the lower portion of the cell spot, its outer edge highly excavated, its inner edge correspondingly rounded—all these spots shining translucent ochreous. Underside, both wings brown strongly washed with vinous. Forewing with the spots as above but white instead of ochreous, the spot in the cell entire. Hindwing with a lengthened subcostal broad streak posteriorly bounded by the subcostal nervure and second subcostal nervule; a discal recurved transverse series of six quadrate spots, of which the two below the posterior end of the subcostal streak are the smallest, a similar but somewhat suffused spot near the base of the wing—the streak and spots all pure silvery. Head and body concolorous with the wings above, palpi and sternum pale ochreous beneath, rest of body and legs concolorous with wings beneath.

Expanse: ♀ 2·0 inches.

Habitat ; Karen Hills, Burma.

A single specimen was obtained in April, 1887, in the Karen Hills by the native collector attached to the Phayre Museum, Rangoon, and I am indebted to Mr. B. Noble for the opportunity of describing it. It is a remarkable species with no near Indian ally, but appears to belong to the same group as the "*Hesperia*" *ornata* of Felder from Buitenzorg, Java, a species which has the spots of the forewing on the underside smaller, and a double series of spots on the hindwing, as shewn in the figure." (*de Nicéville, l. c.*)

48.—PARNARA NAROOA, *MOORE*.

Hesperia narooa, Moore, P. Z. S., 1878, p. 687, pl. xlv, fig. 4.

Parnara narooa, Moore, Lep. Cey., vol. 1, p. 167, pl. 69, figs. 3. *a.b*. (1880-81).

Baoris narooa, Distant, Rhop. Mal., p. 380, pl. lxxxiv, fig. 12 ♀ (1886).

"Male and female. Upperside dark olive brown : forewing with two prominent semi-diaphanous yellowish well-separated spots at end of the cell, a subapical series of three smaller spots and an oblique discal series of five spots: hindwing with two very small discal spots in male, three in female. Underside dark olive brown; forewing marked as above: hindwing with a prominent white spot at upper end of the cell, and a curved discal series of four spots. Body olive green.

Expanse 1 $\frac{9}{13}$ inch.

Hab. Bombay: Ceylon.

Allied to * *H. contigua*, Mabille. Markings similar, but both sexes one-third less in size." (*Moore, P. Z. S. l. c.*)

"Larva pale olive green, delicately speckled with darker green and black dots; a pale unspeckled upper lateral line; head black marked. Pupa pale olivaceous-yellow, darker shaded on thorax." (*Moore, Lep. Cey., l. c.*)

---

* Messrs. Wood-Mason and de Nicéville state they are unable to trace this reference

Also recorded from Andamans and Cachar (*Wood-Mason and de Nicéville*); Calcutta (*de Nicéville*); Poona (*Swinhoe*); Sikkim (*de Nicéville*).

In collection Indian Museum and de Nicéville.

49.—PARNARA PAGANA, *DE NICÉVILLE.*

Parnara pagana, de Nicéville, P. Z. S., 1887, p. 465, pl. xl, fig. 7 ♂.

"Male. Upperside: both wings rich dark brown glossed with purple, the base clothed with long deep ochreous-ferruginous setæ. Cilia ochreous-yellow in the forewing, becoming orange towards the anal angle in the hindwing. Forewing with a spot at the end of the cell sometimes almost quadrate, sometimes constricted in the middle and forming a figure of 8, sometimes quite divided into two spots; three small subapical dots; three increasing discal spots, the anterior one sometimes absent; a spot placed above and against the middle of the submedian nervure, usually round, sometimes oval, rarely entirely absent; all these spots semi-transparent yellow. Hindwing unmarked. Underside: both wings ochreous-brown without any purple gloss, the yellow setæ also absent. Forewing with the base (all except the costa) black; the semi-transparent spots as above, but the one in the submedian interspace developed into a large diffused patch. Hindwing unmarked. Female with the wings a little broader, otherwise exactly as in the male.

Expanse ♂ ♀ 1·9 to 2·2 inches.

Hab. Sikkim.

Nearest to *P. narooa*, Moore, and *P. assamensis*, Wood-Mason and de Nicéville, but differing from both in the hindwing being unspotted.

P. pagana is a common species in Sikkim throughout the year at low elevations." (*de Nicéville, l. c.*). Also recorded from Sikkim by Mr. Elwes.

In collection de Nicéville.

50.—PARNARA PLEBEIA, *DE NICEVILLE.*

Parnara plebeia, de Nicéville, P. Z. S., 1887, p. 466, pl. xl, fig. 2, ♂.

"Male. Upperside: both wings dark brown. Forewing with two or three increasing subapical dots, three increasing discal spots somewhat variable in size and shape; cilia cinereous. Hindwing unmarked; cilia cinereous anteriorly, becoming ochreous towards the anal angle. Underside: both wings paler than above, sometimes tinged with ochreous. Forewing with the semi-transparent spots as above, a diffused large pale patch in the middle of the submedian interspace, sometimes divided into two distinct spots one above the other, sometimes obsolete. Hindwing unmarked. Female: forewing with all the spots larger, always with an additional spot in the submedian interspace and touching that nervure about the middle of its length, often with another smaller spot placed above and beyond the first in the same interspace, these two spots sometimes conjoined. Otherwise as in the male.

Expanse: ♂, ♀, 1·7 to 1·8 inches.
Habitat : Sikkim.

Near to *P. pagana*, de Nicéville, but always smaller, the diaphanous spots white instead of yellow, no spot in the cell, and with no ferruginous setæ at the base of the wings on the upperside. Near to *P. austeni*, Moore, which also occurs commonly in Sikkim, but lacking the two spots in the cell of the forewing, these spots being present also in *P. cahira*, Moore, *P. farri*, Moore, and *P. moolata*, Moore. Nearest of all to *P. kumara*, Moore. From Ceylon specimens of the latter species it differs in the ground-colour of the underside being dull dark brown, sometimes tinged with ochreous, instead of deep ochreous-brown. Mr. Wood-Mason has kindly examined a Ceylon male specimen of the latter species, also a male of *P. plebeia*, and writes regarding them as follow:—" I regard these two specimens as belonging to quite distinct species. The male genital somites and appendages differ very considerably in detail, though identical in plan ; in the Sikkim species the terminal dorsal segment is furnished with a pair of conspicuous conical spines which curve upwards, forwards, and backwards from the disc, and is shorter, and the upper lobe of the claspers is smaller and is embraced at its lower border by the commensurately developed spine of the lower lobe ; while in the Ceylon species the terminal dorsal segment is furnished with shorter spines, from the base of each of which a small cusp is given off backwards, and the sclerite is of greater antero-posterior extent ; and the upper lobe of the claspers is more curved and longer, extending much beyond the spine of the lower lobe ; and the intromittent organ ends in a bilobed spiny brush in the one, and is apparently simple in the other." *P. plebeia* is also near the *P. seriata*, Moore ; the latter, however, is known to me only by the description and figures.

P. plebeia is a common species in Sikkim at low elevations." (*de Nicéville, l. c.*)

Also recorded from Orissa (*Taylor*); Sikkim (*Elwes*).

In collection de Nicéville.

51.—PARNARA KUMARA, *MOORE*.

Hesperia kumara, Moore, P. Z. S., 1878, p. 687.

Baoris kumara, Moore, Lep. Cey., vol. 1, p. 166, pl. 69, figs. 2, 2a. (1881).

" Female. Upperside dark olive brown ; forewing with a transverse discal series of seven yellowish semi-diaphanous spots from before the apex ; cilia pale brownish-yellow. Underside deep ochreous-brown ; forewing marked as above, the lowest spot being yellow : hindwing with a single discal indistinct yellow spot between the two lower median branches. Palpi and body brown ; tarsi ochreous.

Expanse : 1⅜ inches.

Habitat: Canara.

This species is also found in Ceylon, the male differing from the female above and beneath in the first upper and lower discal spots being absent." (*Moore, P. Z. S. l. c.*).

Also recorded from Nilgiris, Mergui (*Moore*); Calcutta (*de Nicéville*); Nilgiris (*Hampson*).

In collections Indian Museum and de Nicéville.

52.—PARNARA SERIATA, *MOORE.*

Hesperia seriata, Moore, P. Z. S., 1878, p. 688.

Baoris seriata, Moore, Lep. Cey., vol. 1, p. 166, pl. 69, figs. 4, 4a. (1881).

" Allied to *H. kumara*. Male differs from the same sex of that species in the discal row of spots being slightly smaller, and in having an additional lower small spot (which is present only in the female of *H. kumara*). Underside greenish-brown; forewing marked as above, the lowest spot suffused.

Expanse: $1\frac{5}{8}$ inches.

Habitat: Ceylon." (*Moore, P. Z. S. l. c.*)

Not in collection Indian Museum, unknown to de Nicéville.

53.—PARNARA MOOLATA, *MOORE.*

Hesperia moolata, Moore, P. Z. S., 1878, p. 843.

Baoris moolata, Distant, Rhop. Mal., p. 379, pl. xxxiv, fig. 10 ♂ (1886).

" Male and female. Upperside dark vinous-brown, slightly olive-brown at the base; cilia cinereous-yellow: forewing with a single small semi-diaphanous white spot at lower end of the cell, two smaller spots obliquely before the apex, and three below on the disk, the lowest being the largest and angular; the female with two spots at the end of the cell, and a small opaque yellowish spot between lower median branch and submedian vein. Underside brighter coloured, marked as above, the yellow discal lower spot in female larger.

Expanse: $1\frac{1}{2}$ inches.

Habitat: Ahsown (2,000 ft.), March. Moolai to Moolat (N. Tenasserim.)

Allied to *H. kumara*, Moore, from S. India, and to *H. cahira* from the Andamans." (*Moore, l. c.*)

In collection de Nicéville.

54.—PARNARA CANARAICA, *MOORE.*

Parnara canaraica, Moore, P. Z. S., 1883, p. 534.

" Male and Female. Upperside dark brown, basal area olive-brown. Male: Forewing with two small oval semi-diaphanous white spots at the end of the cell, three spots obliquely before the apex, and three on the disk: hindwing without makings; cilia brownish-cinereous. Underside paler brown, irrorated with ochreous scales which are thickly disposed along the costa and apex of forewing and across discal area of hindwing: forewing marked as above, also with a small whitish spot above hind margin: hindwing with two discal white spots. Female: forewing with larger spots than in the male, also with a minute dot beneath the lower discal spot and a triangular yellow spot above hind margin; hindwing with three discal semi-diaphanous spots. Underside: forewing as above: hindwing with four discal white spots, and a fifth at end of the cell.

Expanse: ♂ $1\frac{4}{8}$, ♀ $1\frac{7}{8}$ inches.

Habitat: Canara (Ward)." (*Moore, l. c.*)

Not in collection Indian Museum, unknown to de Nicéville.

55.—PARNARA AUSTENI, MOORE.

Baoris austeni, Moore, P. Z. S., 1883, p. 533.

Parnara austeni, Elwes, Trans. Ent. Soc., 1888, p. 448, woodcut n. 3.

"Male and Female. Upperside dark brown. Male: forewing with two small semi-hyaline white spots at end of the cell, two before the apex, and three obliquely on the disk, the two upper of which are small. Female with markings the same, but slightly larger; also with a small yellow spot above the hind margin; cilia cinereous-white. Underside as above; both sexes having also a slight yellowish patch above the hind margin.

Expanse: ♂ $1\frac{3}{8}$, ♀ $1\frac{5}{8}$ inches.

Habitat: Khasia Hills; Cherra Punji.

This species is allied to *H. cahira* from the Andamans and to *H. moolata*; from Tenasserim." (*Moore, l. c.*)

Also recorded from Cachar (*Wood-Mason and de Nicéville*); Sikkim (*Elwes*).

I have numerous specimens of the species from Rangoon.

In collections Indian Museum and de Nicéville.

56.—PARNARA CAHIRA, MOORE.

Hesperia cahira, Moore, P. Z. S., 1877, p. 593, pl. lviii, fig. 8 (♂ only.)

Hesperia cahira, Wood-Mason and de Nicéville, J. A. S. B., vol. l, pt. 2, p. 258, n. 121, ♀ (1881).

"Male. Upperside dark rufous-brown suffused with olive-brown at the base. Forewing with two small yellow spots at end of the cell, two on the disk, and two very small spots before the apex.

Underside rufous-brown.

Expanse: ♂ $1\frac{3}{8}$ inches.

Habitat: S. Andamans (Port Blair).

Near to *H. oceia*, Hewitson; differs in the male above in the absence of the basal tuft of hairs on the hindwing, and beneath also in the absence of the prominent and large white patch and its central spot on the forewing." (*Moore, l. c*). The female described by Moore is really that of *B. oceia*, Hewitson.

"Females have, in addition to the eight spots present in males, a more or less well-developed triangular bright yellow opaque one touching the submedian rather beyond the middle of this, and appearing on the underside as an imperfect band between that vein and the first median veinlet; and, moreover, have the whole underside thickly and evenly clothed with rufous-brown scales." (*Wood-Mason and de Nicéville, l. c.*).

Also recorded from Sikkim (*de Nicéville*); not recorded from Sikkim by Mr. Elwes, but as he considers that *P. cahira* is very probably identical with *P. austeni*, it is not improbable that it is the latter species which is meant by Mr. de Nicéville.

In collection Indian Museum.

57.—PARNARA FARRI, *MOORE*.

Hesperia farri, Moore, P. Z. S., 1878, p. 688.

"Male and female. Upperside ferruginous-brown, base of both wings olive-brown; cilia yellowish-cinereous: forewing with two pale semi-diaphanous spots at end of the cell, and a discal recurved series of seven spots, the four upper and the seventh smallest, the upper three being contiguous and obliquely before the apex, the sixth below end of cell and the largest. Underside greyer brown in the female: forewing marked as in male, except that the lowest spot is more diffused: hindwing without spots.

Expanse : ♂ $1\frac{4}{8}$, ♀ $1\frac{5}{8}$ inches.

Habitat: Calcutta: Cherra Punji." (*Moore, l. c.*)

Also recorded by Mr. de Nicéville from Calcutta.

In collection Indian Museum.

58.—PARNARA TULSI, *DE NICÉVILLE*.

Parnara tulsi, de Nicéville, J. A. S. B., vol. lii, pt. 2, p. 86, n. 30, pl. x, fig. 1, ♂ (1883).

"Male. Upperside rich dark brown with a vinous tinge. Forewing with three very small subapical spots, the middle one out of line, placed nearer the base of wing; an increasing series of three spots outside the cell, placed one each at the bases of the median interspaces, all the spots semi-transparent ochreous-white: the base of the wing and the space below the submedian nervure as well as the base and disc of the hindwing (which is otherwise unmarked) clothed with long ochreous hairs. Underside. Forewing marked as above, but the costa to beyond the middle, and broadly across the disc of the hindwing pale violet-white. Cilia cinereous. No secondary sexual characters. Female differs from the male only in the wings being somewhat broader, and the apex of the forewing less acute.

Expanse: ♂ 1·8, ♀ 1·9 inches.
Habitat : Sikkim." (*de Nicéville, l. c.*)

According to Mr. Elwes, this species is very like *P. plebeia*, above, but easily distinguished by the dull purplish colour of the hindwing below.

In collections Indian Museum and de Nicéville.

59.—PARNARA TOONA, *MOORE*.

Hesperia toona, Moore, P. Z. S., 1878, p. 689, ♂.

Parnara toona, Wood-Mason and de Nicéville, J. A. S. B., vol. lv, pt. 2, p. 383, n. 218 ♀. (1886.)

" Allied to *H. eltola*, Hewitson, Ex. Butt., Hesp., pl. 4, fig. 40.

Male. Differs in the forewing being slightly more elongate, the hindwing more convex exteriorly, and the anal angle less lobed ; markings above similar, those on the forewing narrower, the spot between the median branches elongated and extended to their basal angle ; those on the hindwing very small, the outer spot crossed by a vein. Underside brownish-ochreous ; base of forewing slightly dusky-ochreous ; markings as above.

Expanse : $1\frac{5}{12}$ inches.
Habitat: N.-E. Bengal." (*Moore, l. c.*)

" ♀. Rather larger and less richly-coloured than the male, with the large discal semi-transparent lustrous spot in the first median interspace of the forewing shorter, emarginate, and not extending to the base of the interspace. One example has only the three posterior of the spots of the underside of the hindwing—and these much reduced in size—instead of five, but four are visible on the upperside.

One hundred and twenty specimens [of both sexes] at and round Silcuri [Cachar] from 25th May to 12th August." (*Wood-Mason* and *de Nicéville, l. c.*)

Also recorded from Kumaon (*Doherty*); Nilgiris (*Hampson*) ; Sikkim (*de Nicéville ; Elwes*).

In collections Indian Museum and de Nicéville.

60.—PARNARA ELTOLA, *HEWITSON*.

Hesperia eltola, Hewitson, Ex. Butt., vol. iv, Hesp., pl. iv, fig. 40 (1869).

Parnara eltola, Wood-Mason and de Nicéville, J. A. S. B., vol. lv, pt. 2, p. 384, n. 219, pl. xviii, figs. 6, 6*a*, ♂. (1886).

" Upperside dark brown. Anterior wing with eight transparent white spots, and a spot of yellow near the inner margin : the three largest spots across the middle, two outside of these, and three near the apex. Posterior wing with three transparent spots.

Underside as above, except that it is rufous-brown.

Expanse : $1\frac{13}{20}$ inches.
Habitat: Darjeeling." (*Hewitson, l. c.*)

Also recorded from Kumaon (*Doherty*); Cachar (*Wood-Mason and de Nicéville*); Kangra, N.-W. Himalayas (*Moore*); Sikkim (*de Nicéville; Elwes*).

In collections Indian Museum and de Nicéville.

61.—PARNARA SEMAMORA, *MOORE*.

Hesperia semamora, Moore, P. Z. S., 1865, p. 791.

" Female. Upperside dark vinaceous-brown ; forewing with a curved discal series of five semi-transparent white spots, the two upper minute; hindwing with a broad pure white patch extending half across the wing from abdominal margin. Cilia of forewing brown, of hindwing broadly white. Abdomen with white apex. Underside—forewing dark brown posteriorly, ferruginous-brown anteriorly ; spots as above ; a small suffused patch on the middle of exterior margin, and cilia brownish-white : hindwing with a straight inner bordered ferruginous anterior margin, the rest of the wing pure white : a marginal series of blackish spots ; cilia white. Palpi and legs in front brown. Thorax, abdomen, and legs beneath white.

Expanse: 2 inches.

Habitat: Bengal " (*Moore, l. c.*)

Also recorded from Sikkim (*de Nicéville*).

I caught a single specimen of this species at Beeling, Upper Tenasserim, in April. It differs from the description in having six spots on forewing, the upper three minute; also in a very long series of this species obtained during March and April at Tilin, Yaw District, Upper Burma, I have found invariably three subapical spots.

In collections Indian Museum and de Nicéville.

62.—PARNARA WATSONII, *DE NICÉVILLE*.

Parnara watsonii, de Nicéville, Journ. Bomb. Nat. Hist. Soc., vol. v, p., 223, n. 19 (1890).

Habitat: Upper Burma.

Expanse: ♂, ♀, 1·9 to 2·0 inches.

Description : Male. Upperside, *both wings* rich dark glossy vinaceous-brown. *Forewing* with three subapical rather large conjugated dots ; three increasing discal spots placed obliquely, the second about twice as large as the first, the third about three times as large as the second, the two lower spots with their inner ends convex, their outer concave ; all these spots diaphanous lustrous white ; a somewhat elongated opaque yellowish spot in the submedian interspace placed against the middle of the submedian nervure ; the base of the wing clothed with long dull ochreous hair-like scales. *Cilia* of a slightly lighter shade of colour than the ground. *Hindwing* with the base and abdominal margin thickly clothed with long dull ochreous setæ, the disc with a faint whitish discal macular band. *Cilia* whitish. *Underside*, forewing fuscous, the costa and apex widely ochreous-ferruginous; the diaphanous spots as above; a broad submarginal whitish patch just anterior to the middle of the outer margin of the wing. Hindwing ochreous-ferruginous bearing a large triangular patch of white which occupies all the surface except the costa and the outer margin widely, and a tripartite patch of the ground colour divided by the median nervules in its middle; all the veins that reach the outer margin white. *Head, thorax*, and *abdomen* fuscous above, white beneath, the latter striped with white at the sides. Antennæ black throughout. *Female* almost as in the male, but in two specimens out of three in the forewing there is a fourth smallest discal spot in the lower discoidal interspace, and still another in the submedian interspace almost touching the first median nervule and very small. *Hindwing* with the discal macular white patch more prominent than in the male.

Nearly allied to the "*Hesperia*" *semamora* of Moore,* originally described from Bengal, but occurring in Sikkim, Assam, and Burma from which it differs in all the diaphanous spots of the forewing being larger, the opaque spot in the submedian interspace always present (in *P. semamora* it is found in the females only and is very small), and notably in the large pure white and patch of *P. semamora* being replaced by a whitish shade only in the males, rather more prominent in the females in *P. watsonii*. On the underside of the hindwing there is always a large patch of the ochreous-ferruginous ground-colour in the middle of the white area, this is only occasionally present in *P. semamora* and is when present very small. The markings of *P. semamora* are by no means constant, but the species can always be instantly recognised from *P. watsonii* by the large pure white area on the upperside of the hindwing. In Burma both sexes of the two species have been found flying together, so one cannot be a seasonal form or geographical race of the other.

Described from seven males and three females collected at Tilin, Yaw, Upper Burma, in March, 1890, by Lieutenant E. Y. Watson, after whom I have much pleasure in naming it." (*de Nicéville, l. c.*)

This species occurred very commonly at Tilin throughout March and April in company with *P. semamora*. I also obtained two worn specimens of it in October, 1886, at Poungadaw on the old Burmese frontier about 30 miles north-west of Thayetmyo.

63.—PARNARA PHOLUS, *DE NICÉVILLE.*

Parnara pholus, de Nicéville, Journal, Bombay, Nat. Hist. Soc., vol. iv, p. 172, pl. B, fig. 3 ♀ (1889).

"Habitat: Bhutan.

Expanse: ♂, 2·4; ♀, 2·6 inches.

Description: Male. Upperside, *both wings* rich brown with a shining vinous tinge, the spots translucent rich ochreous. *Forewing* with three conjugated subapical spots, the first a mere dot, the next twice as large,

---

* Proc. Zool. Soc., Lond., 1865, p. 791.

the lowest four times as large as the middle spot; two well-separated spots towards the end of the cell placed inwardly obliquely; a spot about twice the size of these in the middle of the second median interspace, another much larger one completely filling the first median interspace below the innermost spot in the cell, another spot below this at about the middle of the submedian interspace and touching that nervure. *Cilia* concolorous with the wing. *Hindwing* with a small round spot near the end of the cell, three equal-sized spots in a straight line on the disc separated by the second and third median nervules; the base of the wing and the abdominal margin clothed with long yellowish-brown setæ. *Cilia* pale yellow. Underside, *both wings* as above, but the coloration duller. *Forewing* with the spot in the submedian interspace much larger, its edges diffused, pale yellow. *Hindwing* as on the upperside. *Antennæ* and *legs* black throughout; top of *head, thorax* and top of *abdomen* decreasingly clothed with long iridescent bronze-green hairs, thorax below duller. Female, larger than the male. *Forewing* with the lowest subapcial spot larger than in the male, the two spots in the cell conjoined. Otherwise as in the male.

Described from a male obtained by Mr. Wylly, and a female by the native collectors of Mr. Otto Möller, near Buxa, Bhutan. I know of no near ally to this fine species, the largest in the species hitherto described." (*de Nicéville, l. c.*) Mr. de Nicéville also informs me that this species occurs in Sikkim.

64.—PARNARA SARALA, *DE NICÉVILLE*.

Parnara sarala, de Nicéville, Journal, Bombay, Nat. Hist. Soc., vol. iv, p. 173, pl. B, fig. 6, ♀. (1889).

"Habitat: Khasi Hills.

Expanse: ♂, 1·8; ♀, 2·0 inches.

Description: Female. Upperside, *both wings* dark bronzy-fuscous. *Forewing* with a large medially constricted spot at the end of the cell; an elongated spot at the base of the second median interspace; a much larger one towards the base of the first median interspace, its outer end concave, its inner end convex, anteriorly and posteriorly touching the second and first median nervulés; a comma-shaped spot in the submedian interspace, touching the middle of the submedian nervure—all these spots semi-transparent lustrous white; *cilia* fuscous. *Hindwing* with a large oval pale yellow patch on the middle of the disc; and a small patch on the abdominal margin near the base of the wing; *cilia* rich chrome-yellow at the anal angle, gradually shading off into fuscous anteriorly. Underside, *both wings* distinctly glossed with rich purple. *Forewing* with the three discal spots as above, the one in the submedian interspace on the upperside developed into a large outwardly-diffused white patch occupying the middle of the inner margin; a large chrome-yellow quadrate patch above the spot in the cell extending from the subcostal nervure to the costa. *Hindwing* with the oval pale yellow discal patch of the upperside developed into a broad anteriorly-increasing discal chrome-yellow band extending from the abdominal margin to the costa, but with a break between the submedian and internal nervures. *Palpi, thorax,* and *abdomen* above and below clothed with bronzy-green iridiscent hairs; *antennæ* with shaft black, club broken off.

The Rev. Walter A. Hamilton, who obtained the two specimens above described, possesses the wings only of a third specimen placed between tale of what appears to be the male of this species captured in the same locality. In the forewing there are two small well-separated spots in the cell instead of one large one, the two spots below are smaller, the spot in the submedian interspace entirely wanting; otherwise as in the female. This specimen does not apparently possess any secondary sexual characters.

I do not know any near ally to *P. sarala*. The shape of the wings agrees with that of the species of the genus *Parnara*, the probable male, having the forewing less broad, the apex more acute and the outer margin more straight and inwardly oblique than in the female." (*de Nicéville, l. c.*)

65.—PARNARA PARCA, *DE NICÉVILLE.*

Parnara parca, de Nicéville, Journal, Bombay, Nat. Hist. Soc., vol. iv, p. 174, pl. B, fig. 10, ♀. (1889).

"Habitat: Sikkim, Khasi Hills.

Expanse: ♀, 1.9 inches.

Description: Female. Upperside, *both wings* deep vinous-brown. *Forewing,* with three small subapical spots, forming half a circle; two elongated well-separated spots at the end of the discoidal cell; a rhomboidal spot near the middle of the second median interspace; another occupying a similar position in the first median interspace, anteriorly and posteriorly bounded by the second and first median nervules, its inner end well rounded, its outer end convex and the lower corner produced; a rounded spot in the submedian interspace touching that nervule a little beyond its middle—all these spots semi-transparent lustrous white; *cilia* from the inner angle to the second median nervule dull ochreous, anteriorly of the colour of the wing. *Hindwing* with five nearly equal-sized spots forming a rough oval on the disc, the two lowest spots nearer together than the rest; these spots are translucent white in some lights, metallic pale brassy-greenish in others; *cilia* broadly rich chrome-yellow from the anal angle to the termination of the third median nervule, thence to the apex of the wing vinous-brown. Underside, *both wings* of the colour of the upperside. *Forewing* with the spots as on the upperside: the dull ochreous *cilia* of the upperside pale clear yellow, that colour extending a little distance on to the wing. *Hindwing* also with the spots as above;

the chrome-yellow *cilia* of the upperside is pale clear yellow on the underside, that colour extending irregularly on to the wing membrane beyond. Antennæ with the shaft black, becoming ochreous just before the black club; *abdomen* tipped with long chrome-yellow hairs; rest of body, *head and palpi* more or less concolorous with the wings; *femur* and *tibia* of legs black and clothed with very long thick and closely-set black hairs, tarsus anteriorly black, posteriorly deep chrome-yellow, naked.

I place this species but doubtfully in the genus *Parnara*, all the legs being strongly setose, being a character not found in any species of that genus known to me. A somewhat similar character is found in the males only of *Abaratha syrichthus*, Felder, *A. ransonnetii*, Felder, and *A. taylorii*, mihi, which possess a tuft of black hairs over a quarter of an inch in length attached to the coxæ of the front legs and ordinarily lying along the pectus of the butterfly between the middle and hindlegs. These bunches of hairs are probably scent fans and are, moreover, probably susceptible of erection and expansion, but accurate observations on the subject on live specimens are desirable. In describing the genus *Abaratha*,* Mr. Moore states that the legs are naked, this is certainly not the case with the front legs of the males of the type species. Mr. Distant † is also incorrect in saying that the hindlegs of the type species of the genus are strongly pilose; this applies to the fore legs of the male only. It is also quite certain that the species Mr. Distant places in the genus *Abaratha* (*sura*, Moore, and *pygela*, Hewitson), possess a clothing quite different to that setose found in the true *Abarathas*: these species, I think, should be placed in another genus. In the genus *Casyapa*, Kirby, the males have the tibia of the hindlegs extremely hairy.

P. parca is described from a single specimen in my collection obtained by the Rev. Walter A. Hamilton in the Khasi Hills, who possesses the wings of a second example placed between talc from the same region, I also possess another female from Sikkim. I do not know any species at all similarly marked to *P. parca*". (*de Nicéville, l. c.*)

GENUS XVIII.—SUASTUS.

Suastus, Moore, Lep. Cey., vol. i, p. 168 (1881).

"Wings small; forewing elongated, triangular; costa very slightly arched at the base; apex pointed; exterior margin short, oblique, and slightly convex; cell two-thirds the wing, clavate; first, second and third subcostals at equal distances apart, fourth half-way between third and fifth; disco-cellulars inwardly oblique, upper bent inward near the subcostal, upper radial from its angle, lower from their middle; the middle median at one-fifth, lower at three-fifths before end of the cell; submedian straight: hindwing short, broadly oval; apex and exterior margin very convex; cell broad; second subcostal at one-third before

---

* Lep. Ceylon, vol. i, p. 181 (1881).   † Rhop. Malay., p. 390 (1886).

end of the cell; disco-cellulars slightly oblique, radial very indistinct; middle median close to end of the cell, lower at one-third before the end; submedian and internal straight. Body moderate: palpi thick, bristly at the sides, terminal joint long, slender, pointed; antennal club thickish, slender at tip, legs long, slender, almost naked.

Type *S. gremius.*" (*Moore, l. c.*)

66.—SUASTUS GREMIUS, *FABRICIUS*.

Hesperia gremius, Fabricius, Ent. Syst., Sup., p. 433 (1798).
Hesperia divodasa, Moore, Cat. Lep., E. I. C., vol. i, p. 255 (1857).
Suastus gremius, Moore, Lep. Cey., vol. i, p. 168 (1881).

"Upperside dark olive-brown; base of forewing and anal area of hindwing greenish-olive; forewing with three small semi-transparent yellow spots before the apex, two large somewhat quadrate spots and a broken or lunate spot within end of the cell, beneath these are also two lower discal yellow spots—but which are not transparent, the penultimate spot being very small. Underside grey; forewing with the discal area brownish; spots as above: hindwing with a black spot within the cell, and a curved discal series of black spots.

Expanse: $1\frac{1}{2}$ inches." (*Moore, l. c. in Lep. Cey.*)

Recorded from Ceylon (*Wade, Hutchison, Mackwood*); Calcutta (*de Nicéville*); Orissa (*Taylor*); Kumaon (*Doherty*); Kangra District, N.-W. Himalayas (*Moore*); Poona, Bombay, Mhow, Depalpore, Neemuch (*Swinhoe*); Nilgiris (*Hampson*); Sikkim (*de Nicéville*; *Elwes*).

Larva feeds on date-palm.

I have specimens of this species from Madras, Berhampore, Ganjam and Bezvada, Kistna District.

In collection Indian Museum and de Nicéville.

67.—SUASTUS SALA, *HEWITSON*.

Hesperia sala, Hewitson, Trans. Ent. Soc., Lond., ser. 3, vol. ii, 1866, p. 500.
Hesperia sala, Wood-Mason and de Nicéville, J. A. S. B., vol. l, pt. 2, p. 259, n. 124 (1881).

"Upperside dark brown. Anterior wing with four transparent spots; two in the cell minute. Underside grey. Anterior wing dark brown near the base; with three minute brown spots before the apex. Posterior wing with a spot near the middle forming the centre of a circle of several similar brown spots. Expanse, $1\frac{3}{10}$ inches.

Habitat: Singapore." (*Hewitson, l. c.*)

"Female, wings above vandyke-brown with a very faint vinaceous tinge. Anterior wings with four semi-transparent white lustrous spots all close together in middle of the disk, two in the cell, the posterior of which is double the size of the anterior, which is the smallest of all, another immediately behind and in the same straight line with these, quadrate, the largest of all, being fully double the size of the posterior of the two cellular ones, and the fourth placed quincuncially between the second and third and rather smaller than the former of these. Posterior wings immaculate. Wings below paler, darkest around the spots in the anterior ones, all also suffused with vinaceous. Anterior ones with the sutural area ashy, and a large patch in the interno-median area whiter and showing through on the upperside as a somewhat diffused whitish speck. Posterior wings clothed with ashy scales and bearing a cellular roundish spot darker than the ground-colour, around which spot are semi-circularly arranged three or four similar ones, as in *H. divodasa*. Cilia dusky-ashy.

Expanse: 1·23 inches.

Habitat: Andamans. Closely allied to *H. divodasa*, Moore (= *S. gremius*, Fabricius).

The "three minute brown spots before the apex" of the anterior wings mentioned by Hewitson are not discernible in our two specimens; they correspond to the dark brown marks which bound the apical dots on both sides in some specimens of *H. divodasa*, and, consequently, represent the apical semi-transparent dots of that species." (*Wood-Mason and de Nicéville, l. c.*)

68.—SUASTUS ADITUS, *MOORE.*

Suastus aditus, Moore, J. A. S. B., 1884, p. 49.

"Male. Upperside dark violet-brown: forewing with two small quadrate yellow spots at end of the cell, a larger spot immediately beneath end of the cell between the middle and lower medians, and a small spot between the base of upper and middle medians; between the lower median and submedian is a very slight trace of an opaque yellowish streak; cilia edged with grey. Underside paler brown: forewing with the spots as above, and a whitish discal patch below them: hindwing speckled with olive-grey scales, which are most thickly disposed along the abdominal border and form a distinct line along the submedian vein; two dark brown discal spots, one being situated between the middle and the lower medians, the other between the latter and submedian. Body, palpi, and legs beneath olivaceous-grey.

Expanse: $1\tfrac{9}{10}$ inches.

Habitat: Andaman Isles.

Allied to *S. sala* (*Hesperia sala*, Hewitson)." (*Moore, l. c.*)

Also occurs in Orissa (*Taylor*); Nilgiris (*Hampson*); Sikkim (*Elwes*).

Mr. Hampson notes that Nilgiri specimens differ from Andaman ones in the spots of the forewing being smaller, the underside of the hindwing being suffused on the disc with purple.

I have specimens of this species from Beeling, Upper Tenasserim, and from Rangoon.

In collection Indian Museum and de Nicéville.

69.—SUASTUS SUBGRISEA, *MOORE.*

Hesperia subgrisea, Moore, P. Z. S., 1878, p. 689.
Suastus subgrisea, Moore, Lep. Cey., vol. 1, p. 168 (1881).
" Allied to *H. gremius,* Fabricius (Butler, Cat. Lep. Fab. B. M., p. 271, pl., f. 7.)

Male. Differs above in having the three oblique discal spots smaller, the lowest appearing only as a few greenish-grey scales, the upper subapical series of two and a single spot at upper end of the cell very minute. Underside less grey in colour, the forewing marked as above; hindwing with an indistinct black cell-spot and a discal series of spots.

Expanse : 1½ inches.
Habitat: Ceylon." (*Moore, l. c. in P. Z. S.*)
Also recorded from the Nilgiris (*Hampson*).

70.—SUASTUS SWERGA, *DE NICÉVILLE.*

Hesperia, swerga, de Nicéville, J. A. S. B., vol. lii, pt. 2, p. 89, n. 38, pl. x, fig. 12, ♂ (1883).
Suastus mölleri, Moore, J. A. S. B., 1884, p. 49.

" Male. Upperside dark brown. Forewing with a spot at the end of the cell, a larger one below it, and a third (sometimes absent) much smaller, placed outwardly between them at the base of the second median interspace; three increasing conjugated subapical spots (sometimes absent), all these spots semi-transparent lustrous white; a pale ochreous spot placed against and above the middle of the submedian nervure. Cilia slightly paler than the ground-colour of the wing. Hindwing clothed with long pale brown hairs in the middle of the disc. Cilia grey. Underside. Forewing dark brown, the apex widely pale ochreous, this colour decreasing to the inner angle; the spots as above, except that the pale ochreous one placed against the submedian nervure is absent) Hindwing pale ochreous throughout, which is the colour of the cilia on both wings. Body dark brown above, ochreous-white below. The female resembles the male.

The forewing of this species is very long and narrow, and differs in shape from all the hesperids with which I am acquainted.

Expanse: ♂ 1·45 to 1·6, ♀ 1·6 inches.
Habitat: Sikkim." (*de Nicéville, l. c.*)
Also recorded from Sikkim by Mr. Elwes.
I obtained several specimens of this species at Beeling, N. Tenasserim.
In collections Indian Museum and de Nicéville.

GENUS XIX.—SARANGESA.

Sarangesa, Moore, Lep. Cey., vol. 1, p. 176 (1881).

" Wings small; forewing elongated, triangular; exterior margin short, obliquely convex; first subcostal at one-third before end of the cell; discocellulars recurved, upper bent inward at some distance from subcostal, upper radial from its angle, lower from their middle; cell extending nearly two-thirds the wing; middle median at one-sixth and lower at four-sixths before end of the cell, submedian straight; hindwing broadly conical, exterior margin waved; second subcostal at one-fourth before end of the cell; disco-cellulars recurved, radial from their middle; cell extending more than half the wing, broad; middle median near the end, and lower at nearly one-half before end of the cell. Body moderate; palpi compact, finely bristled in front, terminal joint thick, conical, pointed; antennal club lengthened, tip finely pointed.

Type, *S. purendra.*" (*Moore, l. c.*)

71.—SARANGESA PURENDRA, *MOORE.*

Sarangesa purendra, Moore, P. Z. S., 1882, p. 262.

"Upperside greyer than in *S. dasahara*, with paler mottled markings; forewing with a semi-diaphanous white continuous streak across the cell near its end, a continuous small spot above it, three small conjoined spots before the apex, and three on the disc, the middle discal spot large and quadrate. Cilia alternated with grey. Underside paler: forewing marked as above, and with an additional small white lower spot on the disc, and a basal streak below the cell: hindwing with a small whitish spot in middle of the cell, and a less distinct discal curved series.

Expanse: 1¼ inches.

Habitat: Bombay; Umballa; Kussowlie; Kangra" (*Moore, l.c.*)

Also recorded from Hyderabad, Sind (*Swinhoe*); Kumaon (*Doherty*).

In collections Indian Museum and de Nicéville.

72.—SARANGESA DASAHARA, *MOORE.*

Nisioniades dasahara, Moore, P. Z. S., 1865, p. 787.

"Male and female. Upperside, dark olive-brown, with three indistinct suffused blackish transverse bands, the interspaces being slightly grey; forewing with a minute semi-transparent white spot at the extremity of the cell, another immediately above it, and three others curved before the apex. Underside brown; spots and indistinct blackish bands as above, but with the latter on the hindwing somewhat broken, with the interspaces grey, thus giving it a tessellated appearance.

Expanse: 1¼ inches.

Habitat: Bengal." (*Moore, l.c.*)

Also recorded from Kumaon (*Doherty*); Cachar (*Wood-Mason and de Nicéville*); Orissa (*Taylor*); Nilgiris (*Hampson*); Sikkim (*Elwes; de Nicéville*); Kangra, N.-W. Himalayas (*Moore*).

I have obtained this species at Beeling, Upper Tenasserim, at Rangoon, and at Berhampore, Ganjam.

Mr. Taylor also records a second species of *Sarangesa* from Orissa, but gives no description of it.

In collections Indian Museum and de Nicéville.

73.—SARANGESA ALBICILIA, *MOORE.*

Sarangesa albicilia, Moore, Lep. Cey., vol. i, p. 176, pl. 68, figs. 5, 5a (1881).

"Upperside dark vinous olive-brown, with three very indistinct transverse macular fasciæ; forewing with two very minute semi-transparent white spots at end of the cell, another above it, three before the apex, and two less distinct on middle of the disc, all bordered by the macular fascia. Cilia of forewing edged with white; cilia of hindwing entirely white. Underside: forewing as above and sparsely grey speckled: hindwing thickly covered with greyish-white, leaving only the costal border and discal macular fascia brown.

Expanse: ♂ $1 \frac{2}{10}$, ♀ $1 \frac{3}{10}$ inches.

Differs from *S. dasahara* in the hindwing being more prominently white and the cilia also being entirely white." *(Moore, l. c.)*

Habitat: Ceylon (*Wade, Mackwood*); Nilgiris (*Hampson*).

Mr. Hampson notes that Nilgiri specimens differ from Ceylon ones in being dusky instead of white on the underside of the hindwing.

In collections Indian Museum and de Nicéville.

GENUS XX.—TELICOTA.

Telicota, Moore, Lep. Cey., vol. i, p. 169 (1881).
Telicota, Distant, Rhop. Mal., p. 381 (1886) Part.

"Allied to typical *Pamphila* (*P. comma*); forewing more acutely triangular; cell somewhat broader; fourth and fifth subcostal branches and upper radial closer together at their base; disco-cellulars more oblique, upper much longer; middle median at one-fourth and lower at two-fourths before end of the cell (instead of about one-sixth and four-sixths respectively, as in *P. comma*): glandular streak in male more erect, its surface smooth, and extending from submedian to upper median veins; hindwing more produced and lobular at anal angle; second subcostal at one-third before end of the cell, bent downward from its base and in a line with disco-cellulars, which are slightly recurved; radial invisible. Body stout; palpi very compactly clothed; legs more slender, femora and tibiæ much less hairy, hind tibiæ shorter, and tarsi longer; antennal club longer, not so thick, and with a lengthened tip.

Type *T. augias*, Linnæus." *(Moore, l. c.)*

74.—TELICOTA AUGIAS, *LINNÆUS.*

Papilio augias, Linn., Syst. Nat. vol. 1, pt. 2, p. 794, n. 257 (1767).
Pamphila augias, Moore, P. Z. S., 1865, p. 792.
Telicota augias, Distant, Rhop. Malay., p. 382, pl. xxxiv, fig. 23 (1886.)

"Wings above dark ochraceous; anterior wings with the neuration, a broad oblique discal fascia,—commencing at end of cell where it is broadest and ending at submedian nervure,—and a series of outer more or less elongate spots placed between the nervules,—of which the two uppermost are the longest and the third and fourth shortest,—black; the discal fascia has its central area paler in hue; posterior wings with the costal and basal areas,—excluding cell,—a broad scalloped outer marginal fascia widening towards anal angle, abdominal area and the neuration black; fringe of both wings greyish-ochraceous. Anterior wings beneath as above, but the black discal fascia shorter, and the outer marginal spots more or less obsolete; posterior wings beneath with only traces of the black markings above. Body above fuscous, beneath greyish-ochraceous; legs dark ochraceous.

Expanse wings, 30 to 33 millim." (*Distant, l. c.*)

Recorded from the Andamans (*Wood-Mason and de Nicéville*); Calcutta (*de Nicéville*); Bengal (*Moore*); Orissa (*Taylor*); Cachar (*Wood-Mason and de Nicéville*); Myadoung, U. Burmah (*Butler*); Poona, Mhow, (*Swinhoe*); Sikkim (*Elwes*).

I have also obtained this at Berhampore, Ganjam.

Messrs. Wood-Mason and de Nicéville* state that this species may be known from *T. bambusæ* by the golden coloration of the upperside of the forewing extending along the veins to the outer margin, and being throughout paler.

In collections Indian Museum and de Nicéville.

75.—TELICOTA BAMBUSÆ, MOORE.

Pamphila bambusæ, Moore, P. Z. S., 1878, p. 691, pl. xlv, fig. 11, 12, ♂, ♀.

Telicota bambusæ, Moore, Lep. Cey., vol. i, p. 170, pl. 71, fig. 4 (1881.)

Telicota bambusæ, Distant, Rhop. Mal., p. 382, pl. xxxv, fig. 12 (1886.)

"Allied to *P. augias*, Linnæus, from typical Java specimens of which it differs in its somewhat broader and less pointed wings. Markings above similar, but more defined, the borders of the wings blacker, the basal yellow streak on hindwing confined to a terminal spot at the end of the cell, and the abdominal border black. On the underside the markings are also more clearly defined and the interspaces blacker.

Expanse : ♂ 1⅜, ♀ 1⅜ inches.

Habitat: Calcutta. Larva feeds on bamboo." (*Moore, l. c. in P.Z.S.*)

Also recorded from Ceylon (*Wade, Mackwood*); Sikkim (*Elwes*; *de Nicéville*); Kumaon (*Doherty*); Calcutta (*de Nicéville*); Cachar (*Wood-Mason and de Nicéville*); Orissa (*Taylor*); Hatsiega, Upper Tenasserim (*Limborg*); Nilgiris (*Hampson*).

I have obtained this commonly at Berhampore, Ganjam, and in Madras, also rarely at Rangoon.

In collections Indian Museum and de Nicéville.

---

* Journ. A. S. B., vol. lv, pt. 2, p. 384, n. 224, pl. xvii, fig. 1, ♂ (1886).

76.—TELICOTA SIVA, *MOORE*.

Pamphila siva, Moore, P. Z. S., 1878, p. 692.

" Male. Allied to preceding (*P. brahma*). Darker brown above, the base of forewing and lower discal spot only being ochreous, upper semi-diaphanous spots brighter yellow, the one between the lower median branches broader; spots on hindwing prominent. Underside brighter greenish-ochreous, base of costa not ochreous-red; spots on hindwing clear white, very prominent and with dark border.

Expanse : 1¾ inches.

Habitat : Khasia Hills." (*Moore, l. c.*)

77.—TELICOTA BRAHMA, *MOORE*.

Pamphila brahma, Moore, P. Z. S., 1878, p. 691, pl. xlv. fig. 8.

" Male. Upperside deep ochreous-brown, basal area ochreous, cilia cinereous-yellow; forewing with a black-bordered white streak obliquely below the cell, a semi-diaphanous yellowish-white constricted spot at end of cell, a small subapical streak of three spots, and two large discal spots, the one above hind margin being ochreous; hindwing with an indistinct cell spot, and a discal series of three pale yellowish, semi-diaphanous spots. Underside greenish-ochreous; costal base of forewing ochreous-red, hind margin broadly dusky black and centred by a white streak; markings as above but less distinct. Female, brown above, with less ochreous at base, the spots more prominent. Underside as in male.

Expanse : 1⅜ inches.

Habitat : Masuri, 7,000 ft., N.-W. Himalayas.

Nearest allied to *P. subhyalina*, Ménétriés, from China." (*Moore, l. c.*)

GENUS XXI.—PADRAONA.

Padraona, Moore, Lep. Cey., vol. i, p. 170 (1881).

Telicota, Distant, Rhop. Mal., p. 381 (*Part*), (1886).

" Allied to *Pamphila*. Wings shorter, less triangular; no glandular streak on forewing of male. Body smaller, thorax less robust, terminal joint of palpus slender and cylindrical. Antennal club much longer.

Type, *P. mæsa*." (*Moore, l. c.*)

78.—PADRAONA DARA, *KOLLAR*.

Hesperia dara, Kollar, Hügel's Kaschmir, vol. iv, pt. 1, p. 455, n. 4 (1848).

Pamphila mæsa, Moore, P. Z. S., 1865, p. 509, pl. xxv, fig. 9.

" Wings above dusky, a rather broad band on the costa, an oblique spotted fascia, and a squarish spot at the apex of the forewing, yellow; in the middle of the upperside of the hindwing a spot and a transverse band;

underside entirely yellow, chequered with dusky, the inner margin also dusky.

Very like *Hesperia paniscus* and *sylvius* in form and size. The wings brown on the upperside, a broad yellow streak along the costal margin from the base to beyond the middle; not reaching a small almost square spot near which begins an oblique band which runs towards the inner margin, both yellow. The inner margin between the above-mentioned band and the base of the wing is also yellow. The underside of this wing agrees with the upperside except that the whole of the costal margin and apex are also yellow. The hindwings have in the centre a large yellow spot, behind which is a small one of similar colour. Below they are yellow and only brown along the inner margin, the yellow ground-colour is also covered with brown cubical, but not clearly defined, spots. The body above brown clothed with yellowish hairs, the underside, banded with yellow. The antennæ with the club hooked, barred with brown and yellow above.

The butterfly seems to be nearly allied to *Hesperia maro*, Fabricius. Hügel brought a single specimen from the Himalayas.

Expanse: 1 inch." (*Kollar, l. c.*)

Also recorded from Orissa (*Taylor*); Sikkim (*Elwes*); Calcutta, (*de Nicéville*); Kumaon (*Doherty*); Kangra, N.-W. Himalayas (*Moore*); Cachar (*Wood-Mason and de Nicéville*); also as *P. mæsa* from Simla Hills and Kunawur (*Moore*); Mhow (*Swinhoe*); Orissa (*Taylor*).

I have obtained this species commonly at Rangoon.

In collections Indian Museum and de Nicéville.

79.—PADRAONA MÆSOIDES, *BUTLER.*

Pamphila mæsoides, Butler, Trans. Linn. Soc., Lond., second series, vol. i, p. 554 (1879).

Padraona mæsoides, Moore, Lep. Cey., vol. i, p. 171, pl. 71, figs. 5, 5*a* (1881).

Telicota mæsoides, Distant, Rhop. Mal., p. 383, pl. xxxiv, fig. 24 (1886).

"Upperside vinous olive-brown. Male: forewing with a golden-yellow band along base of costa and within the cell, three small connected subapical spots, a narrow sinuous discal band, and a streak along posterior margin; hindwing with a golden-yellow medial discal sinuous band, a less distinct spot within the cell, and a spot on the costa. Cilia yellow. Underside paler, marked as above. Female: differs only in having narrower markings and less prominent costal band on the forewing.

Expanse: ♂ $\frac{9}{10}$, ♀ 1 inch." (*Moore, l. c.*)

Recorded from Ceylon (*Mackwood*); Andamans (*Wood-Mason and de Nicéville*); Kumaon (*Doherty*); Orissa (*Taylor*); Sikkim (*Elwes*).

Mr. Doherty notes that this species differs from *P. dara* "in the rich dark tawny-ochreous colour of the underside, *dara* being greenish-yellow set with dark scales. The markings are almost exactly alike. The prehensores are singularly different: seen from above the uncus of *dara* is gradually acuminate, that of *mæsoides* abruptly truncate and slightly bilobed; seen from the side the uncus of *dara* is slender, tapering and pointed at the tip, that of *mæsoides* which is surmounted by a prominent tuft of hairs, is blunt and rounded at the tip; the clasp of *dara* is much more slender than in *mæsoides*, and its terminal hook much more produced and bent." (*J. A. S. B.*, 1886, p. 139).

Mr. Hampson notes the markings of the underside of hindwing are defined with black.

In collection Indian Museum.

80.—PADRAONA PSEUDOMÆSA, MOORE.

Padraona pseudomæsa, Moore, Lep. Cey., voL 1, p. 170 (1881).

" Allied to *P. mæsoides*. Comparatively larger. Male with paler yellow markings, which are somewhat broader and more disconnected. Female with comparatively narrower and more disconnected markings. Cilia of a paler tint. Underside also paler yellow.

Expanse: ♂, ♀, 1¼ inches." (*Moore, l. c.*)

Recorded from Colombo, Ceylon (*Hutchison*); Orissa (*Taylor*); Nilgiris (*Hampson*).

Of this species Mr. Hampson notes that the underside is ochreous, that of *dara* being greenish.

In collection Indian Museum.

81.—PADRAONA GOLA, MOORE.

Padraona gola, Moore, P. Z. S., 1877, pl. lviii, fig. 9 ♂.

"Male. Upperside dark vinous-brown; forewing with an oblique discal irregular sinuous bordered golden-yellow band, the band bent before the apex and indented at end of the cell; hindwing with a median discal golden-yellow band, and a few hairs of the same colour at the base; cilia edged with golden-yellow. Front of head, palpi and legs golden-yellow. Underside with the bands as above; costa and apex of forewing and the hindwing suffused with yellow; both wings with a blackish streak at end of the cell.

Expanse : 1 inch.

Habitat : S. Andamans (Port Blair)." (*Moore, l. c.*)

Also recorded from Sikkim (*de Nicéville*).

Cachar (*Wood-Mason and de Nicéville*); Orissa (*Taylor*); Nilgiris (*Hampson*).

In collections Indian Museum and de Nicéville.

Mr. Hampson also notes that another species of *Padraona* nearest to *P. gola* occurs in the Nilgiris, of which the fulvous markings occupy the greater part of the upperside of the forewing.

82.—PADRAONA GOLOIDES, *MOORE.*

Padraona goloides, Moore, Lep. Cey., vol. i, p. 171, pl. 71, fig. 3, 3*a* (1880-81).

Telicota goloides, Distant, Rhop. Mal., p. 382, pl. xxxv, fig. 13, ♂ (1886).

"Upperside dark purple-brown. Male: forewing with a golden-yellow oblique discal sinuous band followed by small costal spots before the apex; base of the costa and cell, and two spots at its end also of the same colour; hindwing with a medial discal sinuous golden-yellow band, the hairy scales extending to the base also yellow. Cilia golden-yellow. Underside with less distinct markings as above: costa and apex of forewing and the hindwing suffused with yellow. Body and legs golden-yellow; palpi and front of thorax beneath saffron-yellow. Female: differs only in the discal band being narrow; and not having the yellow costal streak.

Expanse: $1\frac{1}{10}$ inches.

Nearest allied to *P. gola.* Differs from it on both sides in the narrower discal band of the forewing, the band being also disconnected from the costal spots; the band of the hindwing is also narrower." (*Moore, l. c.*)

Recorded from Ceylon (*Wade, Hutchison, Mackwood*). I have what I believe to be this species from Rangoon.

In collection Indian Museum.

83.—PADRAONA PALMARUM, *MOORE.*

Pamphila palmarum, Moore, P.Z.S., 1878, p. 690, pl. xlv, fig. 6, 7, ♂, ♀.

"Differs from typical Java specimens of *P. augias,* Linn. in its much larger size, the male having the markings above paler yellow and broader, the impressed oblique discal streak on the forewing being obsolete, the outer border of the discal band much less sinuous, and the yellow colour not extending along the veins to the exterior margin of the wing.

Female. Paler brown above, the yellow bands on the upperside also paler, less prominent, and there being no yellow along the costal border of forewing, and no median streak from base of the hindwing. Underside pale vinous brown, this colour pervading the upper portion of the discal band on the forewing and the entire discal band on the hindwing.

Expanse: ♂ $1\frac{4}{8}$, ♀ $1\frac{5}{8}$ inches.

Habitat: Calcutta.

Larva feeds on date and cocoanut palms." (*Moore, l. c.*)

Also recorded from Nicobars, Cachar (*Wood-Mason and de Nicéville*); Calcutta (*de Nicéville*).

In collections Indian Museum and de Nicéville.

GENUS XXII.—AMPITTIA.

Ampittia, Moore, Lep. Cey., vol. i, p. 171 (1880-81).
Telicota, Distant, Rhop. Mal., p. 381 (1886), Part.

"Wings shorter, broader and less triangular than in *Pamphila*. Male with a very short glandular streak below the cell. Body narrower; terminal joint of palpi shorter, cylindrical and less pointed. Antennæ with a much slenderer and more pointed club.

Type, *A. maro*." (*Moore, l. c.*)

84.—AMPITTIA MARO, *FABRICIUS*.

Hesperia maro, Fabricius, Ent. Syst. Suppl., p. 432 (1798) ♂.
Cyclopides camertes, Hewitson, Desc. Hesp., p. 43 (1868) ♂.
Ampittia maro, Moore, Lep. Cey., vol. 1, p. 172, pl. 71, fig. 1, 1*a* (1880-81).
Telicota maro, Distant, Rhop. Mal., p. 383, pl. xxxv, fig. 14 ♂, 15 ♀ (1886).

"Upperside purplish olive-brown. Male: forewing with a broad golden-yellow basal band which also crosses the cell, a broad subapical spot, a similar discal spot, and a lower very small spot; a pale band also along posterior margin: hindwing with a broad medial golden-yellow band. Underside: forewing marked as above, and with a marginal row of yellow-bordered blackish dots: hindwing with a subbasal and two discal series of yellow spots. Body, palpi, legs and antennæ beneath yellow. Female with paler similarly disposed but much smaller markings, the basal band on forewing represented only by the termina portion at and above the cell, and the band on hindwing by well separated spots.

Expanse: 1 inch." (*Moore, l. c.*)

Recorded from Ceylon (*Wade, Hutchison, Mackwood*); Orissa (*Taylor*); Cachar (*Wood-Mason and de Nicéville*); Calcutta (*de Nicéville*); Madras (*Swinhoe*); Kangra, N.-W. Himalayas (*Moore*); Nilgiris (*Hampson*).

I have obtained this species commonly at Beeling, Kyaikto, and Sittang in Upper Tenasserim, and also at Pegu and Madras.

In collections Indian Museum and de Nicéville.

85.—AMPITTIA CORAS, *CRAMER*.

Papilio coras, Cramer, Pap. Ex., pl. xxxi, fig. f.

The following description is taken from Cramer's plate:—

Upperside reddish-brown paler at the base of wings with the following yellow spots, one at end of cell with a patch of the basal colour above it, three at apex divided by the veins, two smaller beneath and exterior to these and a more or less maculated band reaching almost to inner margin (this band is not similar on both wings in the plate). Hindwing with a discal row of five yellow spots of which the second is long and divided by a nervure. The underside is stated to be similar. It is described as from Surinam.

Recorded from Bombay (*Swinhoe*).

In collection de Nicéville.

GENUS XXIII.—TARACTROCERA.

Taractrocera, Butler, Cat. Lep. Fab., p. 279 (1869).

Taractrocera, Moore, Lep. Cey., vol. i, p. 172 (1880-81).

"Wings small : forewing triangular : fourth and fifth subcostals, and upper radial from end of the cell; disco-cellulars very oblique, slender, upper longest, lower radial from their middle; the middle median at one-fifth and lower at one-half before end of the cell; submedian recurved; hindwing short, exterior margin very convex; second subcostal at one-fifth before end of the cell; disco-cellulars outwardly oblique, radial from the middle; the middle median near to end and lower at one-third before end of the cell; submedian straight, internal recurved. Body rather stout; palpi erect, terminal joint long and slender; legs naked; antennæ with a short broad spatular club.

Type, *T. mævius.*" (*Moore, l. c.*)

86.—TARACTROCERA MŒVIUS, *FABRICIUS*.

Hesperia mævius, Fabricius, Ent. Syst., vol. iii, p. 352 (1793).

Taractrocera mævius, Butler, Cat. Lep. Fab., p. 279, pl. 3, fig. 13.

Hesperia flaccus, Fabricius, Ent. Syst. Suppl., p. 434 (1798).

Pamphila sagara, Moore, P. Z. S., 1865, p. 792.

Taractrocera mævius, Moore, Lep. Cey., vol. i, p. 172, pl. 70, fig. 5 (1880-81).

"Male and female. Upperside pale olive-brown: forewing with a yellowish-white slender streak at lower end of the cell and joined above at the end by a smaller streak reaching to the costa; three minute spots obliquely before the apex, two below them nearer the outer margin, and a discal sinuous series of small spots: hindwing with a slender medial discal series of very small spots. Cilia yellowish-white. Underside paler; brighter tinted in female; markings more distinct, the end of the discal series of spots on the hindwing extending upward toward the costa, and also having a small spot at end of the cell.

Expanse : $\frac{1\frac{1}{4}}{1\frac{1}{2}}$ inches." (*Moore, l. c. in Lep. Cey.*)

Recorded from Ceylon (*Hutchison, Wade, Mackwood*); Calcutta (*de Nicéville*); Mhow (*Swinhoe*); Kumaon (*Doherty*); N. W. India (*Butler*); Orissa (*Taylor*); Kangra, N.-W. Himalayas (*Moore*); Cachar (*Wood-Mason and de Nicéville*); Nilgiris (*Hampson*); Sikkim (*Elwes*).

I have obtained the species at Beeling (Upper Tenasserim), Poungadaw (Upper Burma), Rangoon, Berhampore (Ganjam), and Madras.

In collections Indian Museum and de Nicéville.

87.—TARACTROCERA CERAMAS, *HEWITSON*.

Cyclopides ceramas, Hewitson, Hesperiidæ, p. 44, n. 10 (1868).

"Alis rufo-fuscis, ciliis latis pallidis, anticis maculis octo fulvis, posticis maculis quattuor: anticis infra apice, posticis omninoochraceis, maculis flavis notatis.

Upperside rufous-brown. Anterior wing with eight minute orange spots: one in the cell, three at the apex, two below these and two nearer the outer margin. Posterior wings with four spots in pairs.

Underside: anterior wing as above, except that the costal margin and apex and whole of the posterior wing are ochreous-yellow. Posterior wing with some indistinct spots of a brighter yellow.

In the collection of W. C. Hewitson.

Also very near to *Rhadama*, but with the yellow spots differently placed.

Expanse: $\frac{18}{10}$ inches.

Habitat: Neilgherries." (*Hewitson, l. c.*)

Also recorded from the Nilgiris by Hampson, and I have obtained in there myself in August. It is readily distinguished from the preceding by the bright yellow colour of the markings which are very pale straw colour in *mævius*.

In collections Indian Museum and de Nicéville.

88.—TARACTROCERA DANNA, *MOORE*.

Pyrgus danna, Moore, Cat. Lep. E. I. C., 1, p. 249.

Pamphila danna, Moore, P.Z.S., 1865, p. 508, pl. xxx, f. 8.

"Upperside dark greenish-brown; forewing with a narrow longitudinal streak within discoidal cell, terminating in two small spots ascending towards the costa, a small geminated spot before the apex, and three irregular shaped spots obliquely on the disc pale yellow; hindwing with a slightly defined longitudinal discoidal streak, a nearly parallel spot, and a transverse series of four discal spots pale yellow; cilia whitish. Underside grey; basal portion of forewing and broadly on hindwing brown-black, the latter with a pale yellowish inner border; base of costa of forewing with spots and marks on both wings, as above but more clearly defined, pale yellow; cilia grey.

Expanse: 1 inch.

Habitat: Simla Hills and Kunawur.

Frequents thistle heads and banks of thyme, at hot mid-day with a rapid, flashing, skipping flight, June." (*Moore, P. Z. S. l. c.*)

I have included this species under *Taractrocera* at the suggestion of Mr. de Nicéville. The other two species of this genus have a weak flight and keep close to the ground usually among long grass.

GENUS XXIV.—CUPITHA.

Cupitha, Moore, J. A. S. B., 1884, p. 47.

" Male. Forewing elongated, triangular, costa arched at the base, exterior margin oblique, posterior margin convex towards the base; first subcostal emitted at nearly one-half before end of cell, the branches at equal distance apart; cell extending to nearly two-thirds length of the wing; disco-cellular almost erect, slightly bent close to upper end and below the middle; the middle median at one-sixth and lower median at four-sixths before end of cell, submedian undulated; on the underside of the forewing is a short, broad, nacreous patch on the middle of posterior margin, across which the submedian is lined with rough scales, and from near the base of the margin projects a broad pencil of long rigid hairs: hindwing short, costa very much arched from the base, apex rounded: costal vein extending to near apex, forked at its base: subcostal bent upward and slightly joined to costal close to the base: subcostal two-branched, first branch from close to end of the cell; disco-cellular very slender, slightly oblique and concave; cell extending to nearly half the wing, of equal width throughout: middle median from near end of the cell, lower at more than one-half before the end, *the portion from the middle median distorted and extending, beneath a drum-like glandular sac, which extends upwards in a circular form within the cell from base of lower median*, the sac or drum as seen from the upperside is flat with a well defined circular rim, and on the underside it stands out from the surface in a corrugated circular form; no radial present; submedian straight, internal vein curved.

Thorax stout; antennæ with a slender club.

Type, *C. tympanifera*." (*Moore, l. c.*)

89.—CUPITHA PURREEA, *MOORE*.

Pamphila purreea, Moore, P.Z.S., 1877, p. 594, pl. lviii, fig. 10.
Pamphila purreea, Wood-Mason and de Nicéville, J.A.S.B., 1881, p. 261.
Cupitha tympanifera, Moore, J.A.S.B., 1884, p. 48.

" Upperside blackish-brown; cilia yellow, slightly alternated with black; forewing with a gamboge-yellow basal streak, and a median oblique irregular band commencing from near apex, extending to hind-margin and terminating at its base; hindwing with a short median yellow band. Underside sulphur-yellow; forewing with a broad dark-brown basal streak, a small spot at end of cell, and a large patch at posterior angle; hindwing with a brown-speckled streak along inner margin, terminating broadly at anal angle. Body above brown, head and thorax interspersed with yellow hairs; abdomen narrowly banded with yellow; palpi black above, yellow below. Legs and body beneath yellow.

Expanse: 1¼ inches.

Habitat: S. Andamans (Port Blair)." (*Moore, l. c.*)

"♀ Larger than the male, with the yellow discal basal throughout in the posterior wings, but only in the interno-median area in the anterior ones, and the yellow portions of the cilia, especially towards the inner and anal angles, darker, inclining to orange.

Expanse: 1·18 inches." (*Wood-Mason and de Nicéville, l.c.*)

Mr. de Nicéville notes that the male has a bare patch at the end of the cell on the upperside of the hindwing on which is placed an oval patch of closely packed scales.

Also recorded from Orissa (*Taylor*); Andamans (*Wood-Mason and de Nicéville*); Nilgiris (*Hampson*); Sikkim (*de Nicéville*).

I obtained two specimens of this at Beeling, Upper Tenasserim, and several specimens since in Upper Burmah. What is universally accepted as one form of the species has been described by Mr. Moore, from Magaree, Pegu, under the name of *C. tympanifera*. According to Mr. Moore it "is a comparatively larger insect that *C. purreea*; the bands on the forewing are broader with more irregular borders, the bands of the hindwing are also broader."

In collections Indian Museum and de Nicéville.

GENUS XXV, *AEROMACHUS*.

Aëromachus, de Nicéville, Journal Bomb. Nat. Hist., Soc., vol. v, p. 214 (1890).

Thanaos, (auctorum nec Boisduval).

"Both wings very small. FOREWING, triangular, *costa* quite straight, *apex* acute, *outer margin* gently convex, *inner angle* rounded, *inner margin* straight, longer than the outer margin; *costal nervure* ending about opposite the apex of the discoidal cell, well separated from the costa, bent upwards to the costa towards its end; base of *second subcostal nervule* nearer to base of first than to base of third, *fourth* subcostal arising very near to base of third, reaching the apex of the wing; terminal portion of *subcostal nervure* (often called a fifth subcostal nervule) ending on the outer margin considerably below the apex of the wing; *upper disco-cellular nervule* short, stout outwardly oblique, straight; *middle* disco-cellular sinuous; lower disco-cellular shorter than the middle, straight, in the same straight line with the middle, inwardly oblique; the *median* nervules with their bases equi-distant, given off very near to the end of the cell, the third median originating at the point where the lower disco-cellular nervule meets the median nervure; the *median nervure* strongly bent upwards from the base of the second median nervule; *submedian* nervure straight. MALE (in the type species only) with a broad oblique stripe of modified scales on the upperside extending from the middle of the submedian nervure to the base of the second median nervule. HINDWING, much rounded throughout; *costa* short; *costal nervure* almost straight, *first subcostal nervule* bent upwards at base, thence straight to apex of wing; *subcostal nervure* strongly bent upwards between the bases of the subcostal nervules, giving the appearance of a third (or upper) disco-cellular nervule, the subcostal nervure and its branches together forming a figure of almost the exact shape of a tuning-fork ; *disco-cellular nervules* outwardly oblique, the *upper* concave, the *lower* shorter than the upper ; the *discoidal* nervule curved, and, like the disco-cellular nervules, very fine but perfectly distinct ; *second median* nervule given off some little distance before the lower end of the discoidal cell, more than twice as far from the base of the first as it is from the base of the third median, all three median nervules, however, arising near to the lower end of the cell ; *submedian* and *internal nervures* straight. Antennæ exactly half the length of the costa of the forewing, with a well-formed club, the tip slightly hooked ; *thorax* rather slender, *abdomen* very slender, FEMALE differs from the male in having the wings broader and more rounded, and lacks in the type species the patch of androconia on the upperside of the forewing. Type, "*Thanaos*" *stigmata*, Moore.

The type of the genus *Thanaos* of Boisduval (1832-33), in which all the species of *Aëromachus* have hitherto been placed, is the "*Papilio*" *tages* of Linnæus, which occurs in Europe and Western Asia (Amurland, &c). *Thanaos* is usually ranked as a synonym of *Nisoniades*, Hübner (1816), of which the type is *bromius*, Stoll, a South American species, which is probably not congeneric with *tages*. *Aëromachus* differs from *T. tages* in the shape of the wings, especially in the hindwing, which in that species is altogether much larger, and has the costa almost straight and very much longer, thus giving quite a different outline to the wing ; the forewing of the male of *T. tages* has the costa folded over on the upperside ; the differences in neuration too are considerable, in the forewing of *T. tages* the first median nervule arises near the bases of the wing, in *Aëromachus* near the lower end of the cell ; and the shape of the discoidal cell of the hind-wing is quite different, in *T. tages* being square-ended, the disco-cellulars being perfectly upright, and of equal length.

The genus *Aëromachus* is, as far as I know, strictly confined to India, where it occurs all along the Himalayas, in Assam, Burma, and again in the hills of South India. They rest with wings closed over the back." (*de Nicéville, l. c.*)

90.—AEROMACHUS INDISTINCTA, *MOORE*.

Thanaos indistincta, Moore, P. Z. S., 1878, p. 694.

Aëromachus indistincta, de Nicéville, Journal Bomb. Nat. Hist. Soc., vol. v, p. 216 (1890).

" Upperside uniform olive-brown ; cilia edged with cinereous.

Underside forewing with a very indistinct grey-speckled submarginal and marginal line : hindwing with indistinct grey-speckled veins, basal interspaces, and two outer indistinct lunular bands. Palpi and body greyish-white beneath.

Expanse: $\frac{8}{10}$ inches.
Habitat: Salween, Moulmein." (*Moore, l. c.*)
Also recorded from Nilgiris (*Hampson*).
I have this species from Rangoon and the Nilgiris.
In collection de Nicéville.

91.—AËROMACHUS OBSOLETA, *MOORE*.

Thanaos obsoleta, Moore, P. Z. S., 1878, p. 694.
Aëromachus obsoleta, de Nicéville, Journal Bomb. Nat. Hist. Soc., vol. v, p. 217 (1890).

"Allied to *T. stigmata*. Differs above in the absence of the short black oblique bar or brand on the forewing; the maculated band being slightly more prominent. Underside similarly speckled, the bands on hindwing not lunular, but composed of a slightly broader series of spots; some spots also present round the cell spot.

Expanse: $1\frac{1}{2}$ inches.
Habitat: Cherra Punji, Assam." (*Moore, l. c.*)
Also recorded from Cachar (*Wood-Mason and de Nicéville*).

Mr. de Nicéville is inclined to believe this species synonymous with the preceding, but as it apparently possesses a discal row of spots on the forewing which is wanting in *A. indistincta*, I have retained them as distinct for the present.

In collection Indian Museum.

92.—AËROMACHUS KALI, *DE NICÉVILLE*.

Thanaos kali, de Nicéville, J.A.S.B., 1885, p. 123, pl. ii, fig. 3, ♂.
Aëromachus kali, de Nicéville, Journal Bomb. Nat. Hist. Soc., vol. v, p. 217 (1890).

"♂ Upperside deep purplish black, the cilia cinereous. Underside slightly paler. Forewing with a discal outwardly angled series of eight pale violet-white dots, an even somewhat larger marginal lunular series. Hindwing with a discal irregular series of pale violet-white spots, within which are some obscure pale markings; a marginal series as in the forewing. Cilia cinereous, marked with dark brown at the ends of the nervules.

Expanse: 1·15 inches.
Habitat: Sikkim (*Otto Möller and de Nicéville*.)
This is a very distinct species." (*de Nicéville, l.c.*)
Also recorded from Sikkim by Mr. Elwes as rare.

Mr. de Nicéville states that this species is easily distinguished on the wing from *T. jhora* and *T. stigmata*, which occur with it, by its much larger size and deep black colour.

In collections Indian Museum and de Nicéville.

93.—AËROMACHUS JHORA, *DE NICÉVILLE.*

Thanaos jhora, de Nicéville, J.A.S.B., 1885, p. 122, pl. ii, fig. 12, ♂.

Aëromachus jhora, de Nicéville, Journal Bomb. Nat. Hist. Soc., vol. v, p. 216 (1890).

"Upperside dark brown; cilia whitish marked with brown at the end of the nervules. Forewing with a discal curved series of about six pale dots. Hindwing unmarked. Underside dark brown, the costa and the apex of the forewing and the entire hindwing greenish-ochreous. Forewing with the discal series of spots as above, and an indistinct marginal lunular series. Hindwing with a very irregular discal series of spots and an obscure marginal series.

Expanse: ·95 to 1·05 inches.

Habitat: Sikkim (*Otto Möller and de Nicéville*).

Nearest to *T. stigmata,* Moore, which occurs commonly in Sikkim with it and is the only species of the genus hitherto described which is furnished with a male sexual mark on the upperside of the forewing." (*de Nicéville, l.c.*)

I have this species from Toungoo in Burmah where it occurs fairly commonly in November.

In collections Indian Museum and de Nicéville.

94.—AËROMACHUS STIGMATA, *MOORE.*

Thanaos stigmata, Moore, P. Z. S., 1878, p. 694.

Aëromachus stigmata, de Nicéville, Journal Bomb. Nat. Hist. Soc., vol. v, p. 216 (1890).

"Male and female. Upperside glossy olive-brown: forewing with a short black bar or brand of raised scales obliquely above middle of hind-margin, and a very indistinct upper discal slightly curved row of six small pale spots: cilia whitish-cinereous, with slight brown bars. Underside paler; costal border of forewing, veins, and basal interspaces of hindwing speckled with greenish-grey; forewing with whitish discal maculated band as above, but more distinct, a spot at end of the cell and a marginal row of lunules less distinct; hindwing with a distinct whitish cell-spot and a sub-marginal and marginal lunular band. Female without the raised bar and the discal band above less distinct.

Expanse: ♂ 1 inch, ♀ $\frac{9}{10}$ inches.

Habitat: Masuri, (7,000 feet), N. W. Himalayas." (*Moore, l. c.*)

Also recorded from Kumaon (*Doherty*); Sikkim (*Elwes*); Kangra, N. W. Himalayas (*Moore*).

In collections Indian Museum and de Nicéville.

GENUS XXVI.—CYCLOPIDES, *HUBNER.*

Cyclopides, Doubleday and Hewitson, Desc. Gen. Di. Lep., p. 520 (1850-52).

"Head as broad as the thorax. Labial palpi remote apart, very hirsute, porrected as long as the head; terminal joint very minute, conical, nearly concealed by the hairs of the preceding joint. Antennæ short with the club stout slightly curved not hooked at the tip, which is obtuse. Wings when at rest erect.

Forewings long, fringe entire, not spotted. Disc dark brown, with orange coloured spots alike in both sexes. The males without a recurved costa or a thickened oblique streak on the disc.

Hindwings short, broad, entire, spotted in the same manner as the forewings.

Hindlegs with the tibiæ destitute of a pair of spurs in the middle.

Abdomen especially in the males, long and slender, with the tip slightly tufted." (*Doubleday, Hewitson, l. c.*)

95.—CYCLOPIDES SUBVITTATUS, MOORE.

Cyclopides subvittatus, Moore, P. Z. S., 1878, p. 692.

C. subvittatus, Wood-Mason and de Nicéville, J.A.S.B., 1886, p. 392, n. 249, pl. xvii, fig. 6, 6*a* (twice life size).

Cyclopides subradiatus, Moore, P.Z.S., 1878, p. 693.

♂ ♀ "Upperside, both wings iridescent vandyke-brown. Forewing with one, two, three, four or five very small pure and bright chrome-yellow streaks divided by the veins placed obliquely beyond the end of the cell; or even immaculate; and with all the cilia lighter yellow than the spots when present and broadly intersected with brown opposite to the end of the veins. Underside, both wings and the bases of the cilia throughout rich vandyke-brown, darker than above and veined and margined with rich chrome-yellow.

Forewing with the costal margin to a little beyond its middle, the costal and subcostal nervules to the costal and outer margins, and the extremity of the third median nervule veined, and the outer margin bordered, with chrome-yellow, so that the wing may be described as increasingly bordered from the base to the apex, and decreasingly from the apex to the sub-median nervure with yellow streaked with dark brown. Hindwing yellow bordered, with the yellow veins broadly edged on both sides with yellow. Antennæ dark brown, ringed and tipped with chrome-yellow. Head thorax and abdomen above dark vandyke-brown, below yellow.

Expanse: ·90 to ·95 inches.

Habitat: Sikkim, where it is not uncommon at low elevations; Bhutan; and Salween, Moulmein, whence the type specimens were received.

The great variation in the number of the small chrome-yellow spots on the upperside of the forewing presented by our specimens from Sikkim and Bhutan suggest at least the suspicion that the *C. subradiatus* of Moore from the Khasia Hills is not specifically distinct from the *C. subvittatus* of the same writer." (*Wood-Mason and de Nicéville, l.c.*)

In the figure referred to above there is a spot in the cell not mentioned in the description.

Also recorded from Kumaon (*Doherty*); Sikkim (*Elwes*).

In collections Indian Museum and de Nicéville.

GENUS XXVII.—HALPE.

Halpe, Moore, P. Z. S., 1878, p. 689.

Halpe, Moore, Lep. Cey., vol. 1, p. 173 (1880-81).

"Allied to *Pamphila* (*P. sylvanus*). Antennæ with a more slender club and longer hook at tip. Forewing shorter; exterior margin more convex; the discal oblique series of raised scales in male shorter and broader. Head and thorax smaller; abdomen slender. Veins similar, the lower median branch being nearer end of the cell." (*Moore, l.c. in Lep. Cey.*).

96.—HALPE BETURIA, *HEWITSON*.

Hesperia beturia, Hewitson, Desc. Hesper., p. 36, n. 31 (1868).

"Alis fuscis; anticis maculis quatuor vitreis: his infra margine costali flavo irrorato, fascia subapicali macularum flavidarum.

Upperside dark brown with four transparent white spots; two near the middle, and two before the apex; the base and the middle of the posterior wing covered with ochreous hair.

Underside paler brown. Both wings with a submarginal band of pale yellow spots. Anterior wing with the costal margin broadly irrorated with yellow. Posterior wing irrorated with yellow near the base: crossed at the middle by a band of yellow spots.

Expanse: 1·4 inches.

Habitat: Nilgiris and Macassar." (*Hewitson, l. c.*)

Also recorded from Calcutta (*de Nicéville*); Andamans (*Wood-Mason and de Nicéville*); Tavoy (*Elwes and de Nicéville*); Nilgiris (*Hampson*).

Messrs. Wood-Mason and de Nicéville note that the number of spots on the forewing vary from six to eight; though only four are mentioned in the description.

I have specimens of the species (named by Mr. de Nicéville) from Rangoon and Berhampore, they have two spots in the cell, three at the apex and two or three on the disc.

In collections Indian Museum and de Nicéville.

97.—HALPE SIKKIMA, *MOORE*.

Halpe sikkima, Moore, P. Z. S., 1882, p. 407.

Halpe sikkima, Elwes, Trans. Ent. Soc., 1888, p. 453, pl. xi, fig. 3, ♂.

Halpe sikkima (*var.*), Elwes. Trans. Ent. Soc., 1888, p. 453, pl. xi, fig. 4, ♂.

"Allied to *H. beturia*. Male. Differs from same sex of that species in the forewing being more acute at the apex, and the exterior margin less convex; the hindwing also is less convex externally, the colour is much darker olivaceous brown. Upperside of forewing with similar spots the two conjoined spots before the apex less obliquely situated, the two discal spots slightly less separated. Underside also darker, the olive yellowish scales uniformly disposed and not forming a marginal macular band or discal fascia on the hindwing. Cilia of both wings brownish white throughout, not alternated with black as in *H. beturia*.

Expanse: 1⅜ inches.
Habitat: Sikkim.

In shape of wings and coloration this species is somewhat like *H. varia* of Japan." (*Moore, l. c.*)

Also recorded from Cachar (*Wood-Mason and de Nicéville*); Sikkim (*Elwes; de Nicéville.*)

Mr. Elwes (*l. c.*), gives a figure of a variety of this species which he thinks may possibly be distinct, it differs in having no spot in the cell, and three instead of two near the apex of the wing.

In collection Indian Museum and de Nicéville.

This species is very closely allied to *H. homolea* Hewitson from Singapore with which it may possibly be identical. The description of *H. homolea* is given* below for reference.

98.—HALPE SEPARATA, *MOORE*.

Halpe separata, Moore, P. Z. S., 1882, p. 407.

Halpe separata, Elwes, Trans. Ent. Soc., 1888, p. 454, pl. xi, fig. 5 ♂, 6 ♀.

"Male. Also allied to *H. beturia*. Forewing comparatively shorter and the hindwing broader; forewing with three conjoined small subapical yellowish-white spots, a transverse spot at upper end of the cell, and two widely separated spots on the disc. Cilia brownish-white, alternated with dark brown. Underside with the costal and outer borders of forewing and entire hindwing covered with golden-olive scales; posterior border of forewing yellow.

Expanse: $1 \frac{8}{10}$ inches.
Habitat: Sikkim." (*Moore, l. c.*)

Also recorded from Kumaon (*Doherty*); Sikkim (*Elwes*); "The female has an additional spot near the hind margin of the forewing not found in the male: the other spots vary in number. It is easily distinguished from other species of the genus by the pale patch on the outer margin of the forewing below." (*Elwes, l. c.*)

In collections Indian Museum and de Nicéville.

---

* *Hesperia homolea*, Hewitson, Desc. Hesp., p. 29: "Upperside dark rufous brown. Anterior wing with five transparent spots; one in the cell two between the median nervures and two before the apex. Underside as above, except that the anterior wing has a submarginal band of ochreous spots, and that the posterior wing has two bands of similar spots, one of which towards the anal angle is more distinct than the rest. Expanse: $1 \frac{7}{10}$ inches. Habitat: Singapore." (*Hewitson, l. c.*)

99.—HALPE KUMARA, *DE NICÉVILLE*.

Halpe kumara, de Nicéville, J. A. S. B., 1885, p. 121, pl. ii, fig. 10 ♂.

" ♂ Upperside deep bronzy-brown. Forewing with five small equal-sized ochreous spots, *viz.*, two conjoined subapical, one at the upper and outer angle of the cell, and two on the disc. Hindwing unmarked. Underside with the costa and apex diminishing towards the inner angle of the forewing and the entire hindwing clothed with deep ochreous scales; the spots of the forewing as above but larger. An anteciliary black line; cilia ochreous, dusky at the end of the nervules in the forewing. Antennæ dusky above, the club and upper portion of the shaft bright ochreous. The sexual mark on the upperside of the forewing indistinct.

Expanse : 1·4 inches.

Habitat : Sikkim (*Otto Möller*).

Allied to *Halpe separata*, Moore, a female specimen of which from Sikkim is before me. Differs from that species in having only two subapical spots, the spot in the cell not transverse, and the posterior border of the forewing on the underside dark brown, not yellow." (*de Nicéville, l. c.*)

In collection de Nicéville.

100.—HALPE AINA, *DE NICÉVILLE*.

Halpe aina, de Nicéville, Journal Bomb. Nat. Soc., vol. v, p. 176, pl. B, fig. 8, ♂ (1890).

" HABITAT : Sikkim.

EXPANSE : ♂, 1·36 to 1·44 inches.

DESCRIPTION : Nearest to *H. kumara*, mihi, of which Mr. Otto Möller possesses eighteen specimens from Sikkim. MALE. UPPERSIDE, *both wings* of a more tawny-ferruginous colour, due to the entire forewing and the basal two-thirds of the hindwing being clothed with a thick coating of long hair-like scales placed upon a deep brown ground. *Forewing* with two conjoined spots in the discoidal cell, the upper spot answering to the single spot of *H. kumara*, the lower spot twice as large as the upper ; three instead of two increasing conjoined subapical spots ; the two discal spots much the same : the " male-mark," however, instead of being a long continuous black streak of modified scales as in *H. kumara* presents the appearance of two obliquely-placed yellow spots exactly as in *H. gupta*, mihi, which can be teazed out by a pin-point into a quantity of fluffy material like down. UNDERSIDE, *both wings* coloured much as in *H. kumara*. *Forewing* with the translucent yellow spots as on the upperside. *Hindwing* unmarked in four specimens, in one specimen with two opaque pale yellow discal spots.

Described from five male specimens in the collection of Mr. Otto Möller, and four in my own." (*de Nicéville, l. c.*)

101.—HALPE GUPTA, *DE NICÉVILLE.*

Halpe gupta, de Nicéville, J. A. S. B., 1886, p. 255, p. xi, fig. 1, ♂.

" Male. Upperside, both wings dark-brown. Forewing with two small spots in the cell placed obliquely one above the other, obsolete in one specimen, two or three conjoined subapical minute spots, two on the disc divided by the second median nervule. Hindwing with some long ochreous hairs in the middle of the disc. Underside, forewing with the costa and apex diminishing towards the anal angle and the whole of the hindwing clothed with greenish-ochreous scales. Forewing marked as above. Hindwing with two or three small pale opaque spots on the disc. Cilia cinereous, tipped with darker at the end of the nervules.

Expanse: ♂ 1·4 to 1·5 inches.

Habitat: Sikkim.

Nearest to *H. kumara*, mihi, differs somewhat in shape, the forewing being narrower and more produced at the apex, the subapical spots smaller. On the underside in *H. gupta* there are some pale spots on the disc of the hindwing, which are absent in *H. kumara*. The shade of the ground-colour is also quite different: in *H. kumara*, it is golden-brown, in *H. gupta*, greenish-ochreous. The sexual mark is rather prominent. Mr. Otto Möller has obtained several male specimens in Sikkim." (*de Nicéville, l. c.*)

The female has not been described. According to Mr. Elwes the male of this species has a double sexual mark on the forewing.

In collection de Nicéville.

102.—HALPE CERATA, *HEWITSON.*

Hesperia cerata, Hewitson, Ent. Mon. Mag., 1876, p. 152.

Hesperia cerata, Hewitson, Desc. Lep. Coll. Atk., p. 4, (1879).

Halpe cerata, Elwes, Trans. Ent. Soc., 1888, p. 454, pl. xi, fig. 8, ♂.

" Upperside dark brown: anterior wing with four transparent white spots; one in the cell sinuated on both sides, two below this between the branches of the median nervure, and one near the apex bifid; posterior wing with a central series of four or five indistinct white spots. Underside as above, except that both wings have a submarginal series of pale spots, that the posterior wing has a white spot near the base, and a transverse central series of six distinct white spots.

Expanse: 1 $\frac{4}{10}$ inches.

Habitat: Darjeeling." (*Hewitson, l. c.*)

Also recorded from Sikkim (*Elwes*); Nilgiris (*Hampson.*)

According to Mr. Elwes, this is the commonest species of the genus found in Sikkim, the female however is rare and differs from the male. If this is the case the female has apparently never been described.

In collections Indian Museum and de Nicéville.

103.—HALPE ZEMA, *HEWITSON.*

Hesperia zema, Hewitson, Ann. and Mag. Nat. Hist., 1877, series iv, vol. xix, p. 77.

Halpe zema, Elwes, Trans. Ent. Soc., 1888, p. 455, pl. xi, fig. 7, ♂.

"Upperside dark rufous-brown. Anterior wing with six transparent white spots, one in the cell, two divided by a branch of the median nervure, and three near the apex: a black linear spot (which denotes the male) from the inner margin. Posterior wing with an indistinct central ochreous spot: the fringe white. Underside as above, except that it is rufous, that the anterior wing has the costal margin and a subapical band ochraceous, and that the posterior wing is crossed from the costal margin to the submedian nervure by a band of pale yellow.

Expanse: 1·3 inches.

Habitat: Darjeeling and Sarawak." (*Hewitson, l.c.*)

Recorded from Sikkim by Mr. de Nicéville as "common, settles on moist spots."

I have obtained a single specimen of the species at Rangoon.

In collections Indian Museum and de Nicéville.

104.—HALPE DOLOPIA, *HEWITSON.*

Hesperia dolopia, Hewitson, Desc. Hesp., p. 27.

Hesperia dolopia, Hewitson, Ex. Butt., vol. v, pl. lv, fig. 60, 61, (1873).

"Upperside dark brown. Anterior wing with six transparent white spots; one in the cell, two between the median nervures, and three before the apex: an indistinct rufous submarginal band. Posterior wing with a central ochreous spot. Underside rufous-brown. Anterior wing as above, except that the inner margin is white, and that there is a submarginal band of pale yellow. Posterior wing with a narrow band from the base, a broad central band, a spot near the anal angle, and a submarginal band pale yellow: a submarginal band of black spots.

Expanse: $1\frac{3}{10}$ inches.

Habitat: North India." (*Hewitson, l. c.*)

Also recorded from Sikkim by Mr. Elwes as rare.

In collections Indian Museum and de Nicéville.

105.—HALPE RADIANS, *MOORE.*

Halpe radians, Moore, P. Z. S., 1878, p. 690, pl. xlv, fig. 1.

"Male. Upperside luteous-brown, basal hairy scales yellow. Cilia cinereous-white: forewing with a pale-yellow constricted spot at end of the cell, and an irregular transverse continuous discal band of spots with their lower angles continued outward along the veins; hindwing with a yellow streak at end of the cell and a short discal band with outer rays. Underside paler, minutely speckled with yellowish-white; forewing as above, the hind margin being also broadly yellow; hindwing with a subbasal spot, all the veins, and two (a median and a discal) transverse sinuous bands pale yellow. Palpi, body beneath and legs yellowish-white.

Expanse: 1½ inches.

Habitat: Dhurmsala, N. W. Himalayas." (*Moore, l. c.*)

In collection Indian Museum.

106.—HALPE SITALA, *DE NICÉVILLE.*

Halpe sitala, de Nicéville, J. A. S. B., 1885, p. 121, pl. ii, fig. 5, ♂.

" ♂. Upperside, forewing dark brown : two minute conjoined subapical dots, two well separated spots placed obliquely near the end of the cell, and two similar ones on the disc, semi-diaphanous ochreous white. The usual sexual mark, somewhat indistinct. Cilia ochreous, dusky at the end of the nervules. Hindwing dark brown with a patch of ochreous hairs in the middle of the wing; cilia ochreous: underside, forewing dark brown, the costa and apex widely ferruginous-ochreous ; the spots as above. Hindwing ferruginous-ochreous ; two conspicuous white dots placed in the median interspaces, two indistinct ochreous spots placed close together between the innermost of the two spots and the anal angle. Antennæ dusky above, the club and upper portion of the shaft below ferruginous.

Expanse: 1.5 inches.

Habitat: Ootacamund, S. India. (*G. F. Hampson*)." (*de Nicéville, l. c.*)

In collections Indian Museum and de Nicéville.

107.—HALPE HONOREI, *DE NICÉVILLE.*

Halpe honorei, de Nicéville, P. Z. S., 1887, p. 464, pl. xl, fig. 8, ♀.

" Female. Upperside ; both wings fuscous. Forewing with the base clothed with yellow hair-like scales, more or less forming streaks between the veins ; a large rhomboidal spot at the outer end of the discoidal cell, two elongated ones, the upper twice the size of the lower, in the median interspaces, two or three subapical conjugated increasing spots, all semi-transparent glistening yellow. Hindwing with all but the costal margin as far as the second subcostal nervule, and the outer margin somewhat narrowly, and the abdominal margin, clothed with long yellow setæ ; a large discal yellow patch beyond the cell divided by the dark nervules and enclosing a blackish dot in the second median interspace. Underside: forewing black all except the costal margin increasingly, the apex widely and the outer margin decreasingly, which are yellowish-ochreous; the semi-transparent spots as above, with two additional somewhat diffused opaque spots placed one above the other near the middle of the submedian interspace, which appear in a somewhat constricted form on the upper-side of one specimen. Hindwing yellowish-ochreous throughout ; a black spot at the end of the cell and about six between the veins outside the cell ; some obscure submarginal blackish spots ; the abdominal margin and a streak in the submedian interspace black.

Expanse : ♀ 1·5 inches.
Habitat : Pulni Hills, S. India.

The markings of this species remind one at once of those of *Plastingia noëmi*, mihi; but there is only one spot in the cell of the forewing, and the yellow in the hindwing is larger in the species now described.

Described from somewhat worn specimens collected by Father D. Honorè, S. J., in the Pulni Hills of S. India." (*de Nicéville, l. c.*)

Also recorded from Nilgiris, (*Hampson*).

In collection de Nicéville.

108.—HALPE DECORATA, *MOORE.*

Halpe decorata, Moore, Lep. Cey., vol. i, p. 173, pl. 71, fig. 2, (1880-81).

" Upperside olive-brown : forewing with a bright yellow streak at end of the cell, the upper part being circular and nearly divided from the elongated lower portion; two smaller spots before the apex, and two slightly larger obliquely-quadrate spots on the disc; base of wing also speckled with olive-yellow scales: hindwing with a medial discal triangular-shaped patch of olive-yellow scales. Underside olive-yellow : forewing with broad brown posterior margin and macular streak bordering the discal spots : hindwing with slender brown-speckled, indistinct, subbasal spots and curved discal macular line. Body palpi and legs beneath yellow; antennæ ochreous-red.

Expanse : 1¼ inches.

" Galle, Morowaka [Ceylon]. Rare" (*Wade*)." (*Moore, l. c.*)

Not in collection Indian Museum and unknown to de Nicéville.

109.—HALPE BRUNNEA, *MOORE.*

Hesperia egena, Felder, Verh. Zool. Bot. Gesch., 1868, p. 284.

Halpe brunnea, Moore, Lep. Cey., vol. i, p. 174, pl. 70, fig. 4, 4*a*, (1880-81).

"Female. Upperside dark vinous-brown; forewing with a very small semi-diaphanous white spot at upper end of the cell, two spots before the apex, and two larger discal spots, the latter deeply concave on their outer border. Cilia pale vinous-brown, alternated with dark brown. Underside as above. Body and palpi beneath pale olive-brown; antennæ black tipped with red.

Expanse : 1 $\frac{4}{10}$ inches.

A single specimen in collection of Captain Wade [from Ceylon]." (*Moore, l. c.*)

In collection de Nicéville.

110.—HALPE CEYLONICA, *MOORE.*

Halpe ceylonica, Moore, P. Z. S., 1878, p. 690, pl. xlv, fig. 9.

Halpe ceylonica, Moore, Lep. Cey., vol. i, p. 173, (1880-81).

" Upperside dark brown ; base of wings and body olive-brown; forewing with two contiguous subapical small white spots, two oblique discal small conical spots with deeply excavated outer border, and a smaller spot at upper end of the cell; the male with an oblique discal streak composed of broad raised darker brown scales; cilia brownish-cinereous, alternated with dark brown. Underside dark brown speckled with ochreous scales; forewing with white spots as above, and a small spot also below the discal series an upper submarginal row of indistinct pale ochreous spots; hindwing with a broad transverse median and maculated discal greyish-ochreous band. Palpi, body beneath and legs greyish-ochreous."

Expanse: ♂ 1 $\frac{4}{10}$, ♀ 1 $\frac{3}{8}$ inches.
Habitat: Ceylon." (*Moore, l. c. in P. Z. S.*)
Also recorded from Orissa (*Taylor*); Nilgiris (*Hampson*).
In collections Indian Museum and de Nicéville.

I have two unnamed species of this genus, one from the Nilgiris and one from Poungadaw, Upper Burma, neither of which Mr. de Nicéville is able to identify, but being only single specimens he has not described them as new.

GENUS XXVIII.—ISOTEINON.

Isoteinon, Felder, Wien. Ent. Monat., vol. vi, p. 30 (1862).

"Genus *Cyclopidi*, Hübn. Led. accedens, antennis multo longioribus, fere ut in *Pterygospideis*, clavatis abdomineque alas posticas haud superante proesertim discrepat." (*Felder, l. c.*) Type, *lamprospilus*, Felder, from Ningpo, China.

111.—ISOTEINON ATKINSONI, *MOORE*.

Isoteinon atkinsoni, Moore, P. Z. S., 1878, p. 693, pl. xlv, fig. 10.

"Upperside dark glossy olive-brown; cilia brownish-cinereous with a brown inner line and indistinct bars: forewing with a small yellow semi-diaphanous spot at end of the cell, three smaller contiguous spots obliquely before the apex, and two contiguous spots obliquely on the disc. Underside speckled with ochreous-green: forewing marked as above; hindwing with a median discal curved series of eight small prominent white spots, and a spot at end of the cell.

Expanse: 1$\frac{1}{10}$ inches.
Habitat: Darjeeling." (*Moore, l. c.*)

I have numerous specimens of this species from Rangoon, so named by Mr. de Nicéville. Mr. Elwes considers this species and the next (*I. subtestaceus*) to be the same. I have specimens of the latter (named by Mr. de Nicéville) from Beeling, Upper Tenasserim, and they differ considerably in the tone of the underside, being of a sort of pale brick red, the underside of the present species being a warm brown dusted with yellowish-green. I can find no difference in the number or arrangement of the spots except that the series on the underside of the hindwing are much more prominent in *atkinsoni*. I have caught this species only in

August and *subtestaceus* only in April, so possible they are seasonal varieties of one another. Mr. Moore's figure of *atkinsoni* agrees very well with my specimens, and Mr. Elwes' figure (Trans. Ent. Soc., 1888, pl. xi, fig. 96), agrees very well with my specimens of *subtestaceus*, except that the series of spots on underside of hindwing have been omitted. Möller also considers the form figured by Elwes to be the spring blood of *I. atkinsoni* which would agree with the dates of my captures noted above. Mr. Elwes also states that *I. khasianus* of which he has seen the type seems to be the present species. In collections Indian Museum and de Nicéville.

112.—ISOTEINON SUBTESTACEUS, *MOORE.*

Isoteinon subtestaceus, Moore, P. Z. S., 1878, p. 844.

" Male and female. Upperside dark olive-brown; cilia cinereous, brown, with a brown inner line and indistinct bars: forewing with a semi-diaphanous spot at end of the cell, a curved series of three subapical contiguous spots, and two larger contiguous spots on the disc. Underside brownish-ochreous, grey-speckled: forewing with the basal area below the cell and the disc fuliginous-black; spots as above; male with a black tuft on hind-margin: hindwing with a curved discal series of six small white spots, and four subbasal spots, the two lower of the latter being contiguous, the third at the end of the cell and the fourth above the cell. Palpi and body beneath ochreous white; legs brownish ochreous, allied to *I. atkinsoni*, from Darjeeling.

Expanse: $1\frac{3}{4}$ inches.

Habitat: Ahsown, (Tenasserim.)" (*Moore, l. c.*)

In collection Indian Museum.

113.—ISOTEINON KHASIANUS, *MOORE.*

Isoteinon khasianus, Moore, P. Z. S., 1878, p. 693.

" Male. Upperside glossy ochreous-brown; cilia brownish-cinereous, with a brown inner line and indistinct bars: forewing with a small silky-white semi-diaphanous spot at end of the cell, three contiguous spots before the apex and two on the disc. Underside rufous-brown; forewing marked as above; a small tuft of black hairs on middle of hind-margin: hindwing grey-speckled, a small white spot at the end of the cell, and a median discal curved series of white dots, each surrounded by dark brown.

Expanse: $1\frac{1}{10}$ inches.

Habitat: Khasia Hills.

A specimen of what I believe to be the female of this species is in the collection of Dr. Staudinger, but without a locality (though probably Indian). It differs from the male above described in having the spots on the forewing slightly larger, the cell spot transversely elongated, and in there being an additional spot below the two on the disc. It is also greyer on the underside; the hind-margin on the forewing has a very pale border; and the hindwing has no perceptible white dots." (*Moore, l. c.*)

114.—ISOTEINON MASURIENSIS, *MOORE*.

Isoteinon masuriensis, Moore, P. Z. S., 1878, p. 693.

"Male and female. Upperside, bluish purple-brown. Cilia white, alternated with brown: forewing with a large semi-diaphanous white quadrate spot at end of the cell, two contiguous elongated spots below it (and in the male a smaller spot below these), three very small spots before the apex. Underside dark brown, numerously specked with ochrey-brown scales: forewing marked as above: hindwing with a discal series of three small indistinct white spots, one being between the subcostal branches the others between the upper median branches. Body and legs beneath speckled with ochreous-green.

Expanse: 1 ⅛ inches.

Habitat: Masuri, N. W. Himalayas." (*Moore, l. c.*)

Also recorded from Sikkim (*Elwes*), and Kumaon (*Doherty*).

In collections Indian Museum and de Nicéville.

115.—ISOTEINON SATWA, *DE NICÉVILLE*.

I. satwa, de Nicéville, J. A. S. B., 1883, p. 86.

"Male. Upperside, rich dark brown. Forewing with two small subapical spots, the lower one twice the size of the upper a rounded spot at the lower outer end of the cell, two similar spots at the base of the median interspaces, the lower one twice the size of the upper, all semi-transparent diaphanous ochreous-white. A small ochreous spot above the submedian nervure touching its middle. Hindwing with the middle of the disc clothed with long greenish-ochreous hairs. Cilia cinereous.

Underside also dark brown but the apex of the forewing and the outer margin of the hindwing broadly washed with purple. Forewing with the spots as above but lacking the one placed against the submedian nervure; the costa to beyond the middle of the wing bears a narrow bright yellow streak widest at its end.

Hindwing with the basal two-thirds also bright yellow, the outer margin of this yellow area very irregular. A small round brown spot near the middle of the cell another above it and one beyond. No secondary sexual characters. Body brown yellow below; antennæ brown above obscurely annulated with yellow below, club brown.

Female differs only from the male in being larger, the wings broader, and the apex of the forewing less acute. There is a second minute spot above the large one in the cell of the forewing.

This is a fairly common species at low elevation below Darjeeling.

Expanse: ♂ 1·3 to 1·4 ♀ 1·55 inches." (*de Nicéville, l. c.*)

Recorded from Kumaon (*Doherty*); Orissa (*Taylor*); Sikkim (*Elwes*), and I obtained two specimens in the Yaw district, Upper Burmah, in March, 1890.

This species is easily distinguished by the colour of the underside, which is somewhat similar to that of *I. cephaloides*.

In collections Indian Museum and de Nicéville.

116.—ISOTEINON CEPHALA, *HEWITSON*.

Hesperia cephala, Hewitson, Ent. Mon. Mag., 1876, p. 152.
Hesperia cephala, Hewitson, Desc. Lep. Coll. Atk., p. 4, (1879.)
Isoteinon cephala, Elwes, Trans. Ent. Soc., 1888, p. 456, pl. xi, fig. 10, ♂.

" Upperside, dark brown, the fringe brown and white alternately; anterior wing with three transparent white spots and an opaque spot near the inner margin—one at the middle bifid, one at the apex trifid, and one below it; posterior wings with two transparent spots near the middle.

Underside, anterior wing as above, except that the costal margin from the base to the transparent spot, and the outer-margin from the apex to the middle, are yellow; posterior wing yellow, with a black spot near the base, a third white spot adjoining the transparent white spots, which are bordered below with rufous-brown, the outer-margin rufous-brown.

Expanse: $1\frac{7}{20}$ inches.

Habitat: Darjeeling." (*Hewitson, l. c.*)

Also recorded from Sikkim (*de Nicéville*).

I have obtained this species at Beeling, Upper Tenasserim in April, and very commonly at Tilin in the Yaw district, Upper Burmah, during March and April.

In collections Indian Museum and de Nicéville.

117.—ISOTEINON CEPHALOIDES, *DE NICÉVILLE*.

Hesperia? cephaloides, de Nicéville, J. A. S. B., 1888, p. 288, pl. xiii, fig. 4, ♂.

" Male. Upperside, both wings dark purplish-brown; cilia alternately black and white. Forewing with a large somewhat square spot at the end of the cell, a little larger rhomboidal one below it in the first median interspace, and a much smaller square one at the middle of the second median interspace, three small conjoined round subapical dots, of which the upper one is the largest, the middle one the smallest, all translucent white; an opaque dot touching the submedian nervure in the middle of the submedian interspace. Hindwing with three translucent white spots forming an equilateral triangle, of which the two at the base are largest and equal, and the apical one is a mere dot. Underside forewing with a broad costal streak occupying the upper half of the discoidal cell and reaching to just beyond the middle of the wing, and an apical patch, bright chrome-yellow, between which streak and patch the ground is castaneous, the rest of the wing black; the translucent white spots as on the upperside, but with two additional small black spots between the lowest of the subapical series and the spot in the second median interspace. Hindwing with the basal half of the wing chrome-yellow, the outer half castaneous; a small round castaneous spot near the base of the wing, the three discal translucent spots as on the upperside, but with two additional opaque round spots, one near the costa at the inner edge of the castaneous portion of the wing, the other in the middle of the submedian interspace, all five spots surrounded by a fine black line; these are traces of a series of blackish spots between the veins near the margin. Head, thorax, and abdomen black above, beneath, legs, and palpi chrome-yellow.

Expanse: ♂ 1·6 inches.
Habitat: Karen Hills, Burmah.

Very near to, but quite distinct from, *Hesperia cephala*, Hewitson, a fairly common Sikkim species, from which it differs in its large size, and in the following particulars:—the subapical series of spots on the forewing has the middle spot the smallest and the upper one the largest, while in *H. cephala*, the series is an increasing one; in *H. cephala* the spot below these is in the lower discoidal interspace; on the hindwing in *H. cephaloides* there are three small spots, in *H. cephala* there are two only, both large, the outer one very large; on the underside in *H. cephala* the costal yellow streak extends uninterruptedly from the base to the apex, in *H. cephaloides* it is interrupted by a large castaneous patch; in *H. cephala* the hindwing is entirely yellow, in *H. cephaloides* the basal half only is yellow, the outer half being castaneous; the spots too are very different and in greater number and occupy different positions. I am indebted to Mr. B. Noble, the curator of the Phayre Museum, Rangoon, for the opportunity of describing this interesting species of which he has obtained two specimens. They were captured by the native collector attached to that institution in the Karen Hills in April 1887." (*de Nicéville, l. c.*)

In collection de Nicéville.

118.—ISOTEINON PANDITA, *DE NICÉVILLE*.

Isoteinon pandita, de Nicéville, J. A. S. B., 1885, p. 121, pl. ii, fig. 14, ♂.

" ♂. Upperside brown, sparsely clothed with ochreous scales. Forewing with a quadrate transverse spot at the end of the cell, three conjoined subapical ones, and two similar discal ones, semi-diaphanous ochreous. Cilia ochreous. Hindwing immaculate. Underside brown, forewing with the apex widely, and the costa and the entire hindwing ferruginous ochreous. Forewing with the spots as above. Hindwing with a very indistinct small black spot at the end of the cell, and a discal series of similar short streaks between the nervules. Cilia ochreous. Antennæ black, the tip of the club and the upper portion of the shaft below the club ferruginous.

Expanse: 1·2 inches.
Habitat: Sikkim. (*Otto Möller*)." (*de Nicéville, l. c.*)
In collection de Nicéville.

119.—ISOTEINON FLAVIPENNIS, *DE NICÉVILLE*.

Isoteinon flavipennis, de Nicéville, J. A. S. B., 1885, p. 122, pl. ii, fig. 4, ♀.

" ♂ and ♀. Upperside brown glossed with purple on the outer area of the forewing, the hindwing bearing a patch of long ochreous hairs in the middle of the disc and on the abdominal margin. Forewing with the following white semi-transparent spots:—a small round one in the middle of the upper discoidal interspace; two at the end of the cell, one above the other the lower twice the size of the upper; a large spot in the interspace below and a small one placed in the second median interspace near its base. Underside with the costa and apex widely and the entire hindwing ferruginous-ochreous glossed and marbled with purple, the disc of the forewing dark brown, the inner margin paler. Forewing with the spots as above; hindwing with a dark brown spot in the cell and a series of five or six similar spots placed around the cell. Cilia cinereous, dark brown at the end of the nervules. Antennæ black annulated with pale ochreous beneath, the club, all except the extreme tip, pale ochreous beneath.

Expanse: 1·3 to 1·4 inches.

Habitat: Buxa, Bhutan (*Moti Ram*); Sikkim (*Otto Möller*); and South Andaman Island (*A. de Rocpstorff*)." (*de Nicéville, l.c.*)

In collections Indian Museum and de Nicéville.

120.—ISOTEINON MICROSTICTUM, *WOOD-MASON and DE NICÉVILLE.*

Isoteinon microstictum, Wood-Mason and de Nicéville, J. A. S. B., 1886, p. 385, pl. xvii, fig. 3, ♂, fig. 3a. ♀.

" ♂. Upperside, both wings dark vandyke-brown, suffused with purple, especially on the costal and outer margins and the veins; the cilia ochreous-grey. Forewing with five small semi-transparent white lustrous spots, two (the first of which is very minute) before the apex, and three discal, one geminated in the cell consisting of an anteriorly outwardly convex thin crescentic and a posterior triangular portion, another about the same size behind and a little external to this between the first and second median nervules sub-crescentic in shape, with its convexity turned towards the base and a third squarish external to, and in front of, and rather less than half the size of, this again between the second and third median nervules. Underside, forewing lighter than above, the translucent spots as on the upperside; with an indistinct submarginal band of spots darker than the ground; a dark anteciliary line; the cilia obsoletely intersected at the veins with dark, and an indistinct whitey-brown spot, touching the first median nervule and the satiny patch, extending from the base nearly to the outer angle and from the interno-median fold to the inner margin of the wing, which bears a conspicuous fringe of slate-grey setæ in part projecting straight backwards and outwards from the edge and in part turned up so as to spread out fanwise over the satiny ashy patch. Hindwing darker and more suffused with purple than the forewing, with a dark anteciliary line, but even less distinctly intersected cilia; with some dark mottling indistinctly arranged in three bands, one subbasal and two closer together discal or submarginal, and with an indistinct dot between the costal and subcostal nervures, another near the end of the cell, on one side only in one specimen and on both sides in the other, a third between the second and third median nervules in one specimen, and a fourth between the first and second in the other all ochreous-white.

♀. Upperside, both wings much lighter than in the male. Forewing with the spots much larger and more numerous, there being an additional subapical one, a very minute dot just in front of the third median nervule, both semi-transparent and a third opaque yellow one (present in one male), touching the submedian nervure in front, and the third discal spot being quadrate with the inner and outer ends roundly emarginate. Underside both wings with the dark markings more distinct than in the male. Forewing devoid of the ashy patch and fringe of setæ seen in the opposite sex.

Hindwing with two dots behind the costal nervure instead of one, one in the cell, and another between the second and third median nervules on one side only so minute as to be scarcely discernible.

Expanse : ♂ 1·4, ♀ 1·5 inches.

Allied to *I. flavipennis*, de Nicéville, but differs in markings and notably in the colour of the ground of the underside, which is pale vandyke brown in *I. microstictum* and ferruginous ochreous in *I. flavipennis*.

Two males 26th and 27th May, and one female 28th May, Silcuri, [Cachar]." (*Wood-mason and de Nicéville, l. c.*)

In collection Indian Museum.

121.—ISOTEINON FLAVALUM, *DE NICÉVILLE*.

Isoteinon flavalum, de Nicéville, P. Z.S., 1887, p. 463, pl. xi, fig. 10, ♂.

"Male. Upperside: both wings dark brown. Forewing with three small subapical increasing spots, the upper one minute; a small quadrate spot at the lower outer end of the cell, an elongated one at the base of the second median interspace, a much larger quadrate one below it and placed nearer the base of the wing in the first median interspace, all semi-transparent diaphanous ochreous. Hindwing with the middle of the disc clothed with long greenish-ochreous hairs. Cilia cinereous throughout. Underside: forewing also dark brown, the spots as above, the costa narrowly and the apex widely (but not reaching the anal angle or the outer margin) yellow. Hindwing yellow throughout except the outer margin, which is increasingly dark brown, widening to the anal angle; a conspicuous though small black spot on the discoidal cell, three small dark-brown ring-spots placed very close together below it, and a fourth minute black spot well separated from the others towards the apex. No secondary male sexual characters. Head and body above dark brown, below with legs yellow. Antennae black, the club tipped beneath with white.

Expanse: ♂ 1·2 inches.

Habitat: Sikkim.

This pretty and very distinct little species is nearest allied to *Isoteinon satwa*, mihi, but is abundantly distinct; the underside has no purple washing, and on the hindwing the yellow coloration occupies nearly the entire surface; in *I. satwa* it is confined to the anterior half of the wing." (*de Nicéville, l. c.*)

Type in collection Möller.

122.—ISOTEINON VINDHIANA, *MOORE*.

Isoteinon vindhiana, Moore, P. Z. S., 1883, p. 533.

" Male. Upperside dark olive-brown; cilia cinereous; forewing with a small yellow semi-transparent spot at upper end of the cell, three conjoined subapical spots, two discal spots, and a small oval spot above the submedian vein. Underside dusky ochreous: forewing with the posterior area broadly black; spots as above; hindwing with a yellow lunule at end of the cell, a small spot above it and five discal spots.

Expanse: $1\frac{2}{12}$ inches.

Habitat: Jubbulpore (*Span*)." (*Moore, l. c.*)

Also recorded from the Nilgiris by Mr. Hampson, who considers this and the two succeeding species to be identical, *vindhiana* being the dry season form, *nilgiriana* the wet season form, and *modesta* which was described from a single specimen obtained by Mr. Lindsay, a variety.

In collection Indian Museum and de Nicéville.

123.—ISOTEINON NILGIRIANA, *MOORE*.

Isoteinon nilgiriana, Moore, P. Z. S., 1883, p. 533.

" Male. Allied to *I. vindhiana*: forewing with similarly disposed spots, which differ in being white, somewhat smaller, narrower, and the subapical conjoined spots disposed in a smaller row; the spot above submedian obsolescent. Underside uniformly ochreous-brown: forewing with the spots as above, the submedian obsolete; hindwing with a small dusky-black spot at end of the cell, and a dusky-black discal row of spots.

Expanse: $1\frac{2}{12}$ inches.

Habitat: Coonoor, Nilgiris (*Lindsay*)." (*Moore, l. c.*)

Also recorded from Mhow, and Matheran (*Swinhoe*); and Nilgiris (*Hampson*).

In collection de Nicéville.

124.—ISOTEINON MODESTA, *MOORE*.

Isoteinon modesta, Moore, P. Z. S., 1883, p. 534.

" Female. Allied to *I. nilgiriana*: forewing narrower and less triangular in shape, with a minute very indistinct spot at upper end of the cell, two similar minute subapical spots, and two discal spots.

Underside brownish-ochreous, grey-speckled; forewing with spots as above; hindwing immaculate.

Expanse: $1\frac{1}{4}$ inches.

Habitat: Coonoor, Nilgiris (*Lindsay*)." (*Moore, l. c.*)

125.—ISOTEINON FLEXILIS, SWINHOE.

Isoteinon flexilis, Swinhoe, P. Z. S., 1885, p. 147.

" ♂ ♀. Upperside dark shining olive-brown; cilia pure white: forewing with two small semi-diaphanous spots, one at the upper end of cell and one above it: three contiguous subapical spots, the top spot very minute, and in the male sometimes absent; another outer very minute dot, which also is often absent in the male; and three larger spots obliquely—two on the disc and one touching the submedian nervure; hindwing unmarked.

Underside paler spots as above; forewing with a blackish longitudinal shade covering the lower half of the wing; hindwing with an indistinct diffused discal fascia of same colour.

Expanse: $1\frac{1}{10}$ inches.

Habitat: Poona, December." (*Swinhoe, l. c.*)

Not in collection Indian Museum and not known to de Nicéville.

126.—ISOTEINON MASONI, MOORE.

Pamphila masoni, Moore, P. Z. S., 1878, p. 842, pl. 411, fig. 5, ♂.
. *Isoteinon masoni*, Elwes and de Nicéville, J. A. S. B., 1886, p. 442, pl. xx, fig. 4, ♂.

" ♂. Upperside both wings dark brown. Forewing with two conjoined spots placed obliquely near the end of the cell, two subapical dots, the lower four times the size of the upper, a quadrate spot in the second discal interspace, two spots in continuation in the submedian interspace, all bright ochreous. Cilia cinereous. Hindwing with a small patch of yellow in the middle of the disc. Cilia ochreous. Underside forewing with the costa narrowly, the apex widely, decreasing rapidly to the angle, bright ochreous, the rest of the wing black; the spots as above, but the lower of the two in the submedian interspace much larger and diffused, two short dark streaks placed outwardly against the subapical dots, beyond which is a submarginal series of obscure spots of a paler yellow than the ground on which they are placed. Cilia dark brown. Hindwing bright ochreous, with an obscure discal series of dark spots of which one in the upper subcostal interspace is alone prominent. Cilia ochreous defined inwardly by a fine dark brown line. Antennæ with the shaft above fuscous, the anterior half of the club ochreous, the anterior half of the shaft below ochreous. No secondary sexual characters.

Expanse: ♂ 1·15 inches.

Marked almost exactly as in *Halpe honorei*, de Nicéville (which however may be an *Isoteinon*, the male being unknown), from South India but all the spots on the forewing smaller, the one in the discoidal cell nearly divided into two portions, the discal patch on the hindwing less than half the size and the insect itself smaller.

A single specimen only was obtained from Tavoy.

Judging from the figure alone it appears to us that Moore's type specimen was a female, he described it, however, as a male. It is considerably larger than our male." (*Elwes and de Nicéville, l. c.*)

This species was originally described by Mr. Moore from a specimen obtained by Limborg at Hatsiega, Upper Tenasserim, and he gives the expanse as $1\frac{4}{10}$ inches, otherwise his description is substantially the same as that quoted but not so detailed.

I caught a few specimens at Tilin, Yaw District, Upper Burmah, during March, 1890.

In collection Indian Museum.

127.—ISOTEINON INDRASANA, *ELWES and DE NICÉVILLE.*

Isoteinon indrasana, Elwes and de Nicéville, J. A. S. B., 1886, p. 441, pl. xx, fig. 5 ♀.

" ♀ Upperside, both wings brown with a ferruginous gloss. Forewing with a small round subapical dot, a similar spot but twice the size at the lower outer end of the discoidal cell, a slightly larger triangular one in the second median interspace, and a large quadrate one in the first median interspace, all semi-transparent yellow; an elongated spot touching the middle of the submedian nervure in the submedian interspace opaque yellow. Cilia cinereous. Hindwing with an obscure yellow patch on the middle of the disc. Cilia ochreous. Underside, forewing with the costal broadly and the apical half of the wing decreasing to a point at the anal angle dark ochreous, the rest of the wing fuscous; the spots as above. Cilia fuscous, becoming ochreous towards the anal angle. Hindwing dark ochreous throughout, the veins a little paler. Cilia ochreous. Antennæ wanting. Head and body above dark brown, beneath ochreous.

Expanse: ♀, 1·35 inches.

A single example from Tavoy. We do not know any species to which *I. indrasana* is nearly allied." (*Elwes and de Nicéville, l. c.*)

In collection Indian Museum.

128.—ISOTEINON IAPIS, *DE NICÉVILLE.*

Isoteinon iapis, de Nicéville, Journ. Bomb. Nat. Hist. Soc., 1890, vol. v, no. 3, p. 213, pl. E, fig. 9, ♂.

" Habitat: Burmah, Malay Peninsula.

Expanse: ♂ 1·5 inches.

Description: Male. Upperside, *both wings* dark shining brown, becoming of a deeper shade towards the outer margins. *Forewing* with the following semi-transparent lustrous yellowish spots:—two towards the end of the discoidal cell, the upper one a round dot, the lower larger, elongated, comma-shaped; a pair of conjoined subapical dots divided by the terminal portion of the subcostal nervure (which in the *Hesperiidæ* appears always to end on the outer margin some little distance below the apex), the lower a little the larger; three discal spots placed obliquely, the upper one in the lower discoidal interspace a mere dot, equal in size to the lower subapical dot, the middle spot in the upper median interspace about four times as large, its outer edge concave, its inner convex, the lowest spot about four times as large as and shaped like the spot above, placed in the first median interspace. *Hindwing* unmarked, but the abdominal margin and especially the anal angle fringed with very long hairs. Underside, *both wings* much paler than above, sprinkled throughout thickly with dull ochreous scales. *Forewing* with the inner margin broadly pale yellow, a tuft of long black hairs attached to the margin. *Hindwing* with a discoidal and four or five discal small very obscure dark spots, which appear to be formed by a portion of the ground-colour being left free from the dull ochreous scales. *Antennæ* above entirely fuscous, beneath with an increasing yellowish streak towards the apex, not extending to the extreme tip. *Cilia* of the forewing concolorous, of the hindwing cinereous.

Apparently nearest to *I. subtestaceus*, Moore, of which there is one of the original specimens from Upper Tenasserim taken by Mr. Ossian Limborg in 1876-77 in the collection of the Indian Museum, Calcutta. Differs therefrom in having two spots instead of one in the discoidal cell of the forewing, two instead of three subapical dots, three instead of two discal spots, which latter also differ widely in size, shape, and position. The coloration of the underside is also quite different.

The type specimen of *I. iapis* is from Johore in the Malay Peninsula, and was kindly forwarded to me by Mr. W. Davison. In the Indian Museum, Calcutta, are two small specimens taken by Dr. J. Anderson in the Mergui Archipelago on 11th December, 1881, and 11th April, 1882, respectively, which are undoubtedly the same species, though too worn to be identified by Mr. Moore when working out the collection of which these specimens formed a part. On one of them Mr. Moore placed a ticket on which is written "not *moolata*," which is, however, a *Parnara*, and not an *Isoteinon*." (de Nicéville, l. c.)

GENUS XXX.—SATARUPA.

Satarupa, Moore, P. Z. S., 1865, p. 780.

"Palpi stout densely pilose, erect, projecting in front of the head; third joint minute, conical. Antennæ moderate. Body very stout. Legs slender; femora slightly pilose beneath; hind tibiæ pilose at the side and beneath; middle tibiæ with a pair and hind tibiæ with two pairs of apical spurs. Wings—forewing acute; costa nearly straight, exterior margin oblique; hindwing rounded exteriorly in the male, angled at the apex, and in the middle of exterior margin of the female." (*Moore, l. c.*)

Daimio bhagava, Moore 129.—SATARUPA BHAGAVA, *MOORE*.

Satarupa bhagava, Moore, P. Z. S., 1865, p. 781.

"Upperside dark olive-brown: forewing with a triangular series of three discal semi-transparent white spots, the first being large and within the extremity of the cell, the second quadrate and beneath the first, the third exterior to their juncture; beneath these are small black spots bordering a brownish white streak from middle of posterior margin, a recurved series of small similar white spots before the apex; hindwing with a broad brownish white subbasal transverse band, bordered by a semi-circular series of black spots, those exteriorly assuming the form of streaks between the veins. Abdomen with a white band. Underside as above. Palpi and thorax in front beneath orange-yellow. Cilia brown.

Expanse: 1¾ inches.

Habitat: N. E. Bengal." (*Moore, l. c.*)

Also recorded from Orissa (*Taylor*); Sikkim (*de Nicéville; Elwes.*); Margherita, Assam (*Doherty*).

Mr. Elwes only doubtfully records this species, as he considers that the names *bhagava, phisara*, and *narada* really include only two species, one with narrower yellowish band on hindwing which he considers to be *phisara*, and one with pure white and broader bands which he considers to be *narada*.

I have obtained this species at Rangoon and at Poungadaw, U. Burmah.

In collections Indian Museum and de Nicéville.

The following local form has been described from the Andamans.

Tagiades bhagava, var. *andamanica*, Wood-Mason and de Nicéville, J. A. S. B., 1881, p. 256, pl. iv, fig. 5, ♂.

"Male. With the cream-coloured subbasal band of the posterior wings in one specimen narrower and not continued on to the anterior wings, in another as broad as in an Upper Tenasserim example and continued faintly on to the anterior wings: with the spot at the end of the cell larger than in the female and not isolated from the fuscous outer margin: and with the transverse abdominal band concolorous with the subbasal.

Female. Wings above paler with the spots of the anterior wings whiter and larger and the band of the posterior ones pure white, much broader and extending on to the anterior ones broadly up to the submedian vein and thence narrowly up to the first median veinlet between the two pairs of black spots.

Posterior wings with a black speck at the end of the cell on a white ground on both sides and the two anterior of the semi-circular series of black spots on the upperside nearly but on the underside wholly placed on the white subbasal band.

A specimen from the Sikkim hills 3,000 feet, differs in having the band broader both on the posterior wings and between the two rows of spots in the anterior ones.

Habitat: S. Andaman. Sikkim?" (*Wood-Mason and de Nicéville, l. c.*)
In collection Indian Museum.

130.—SATARUPA SAMBARA, MOORE.

Goniloba sambara, Moore, Cat. Lep. Mus. E. I, C., i, p. 246.
Satarupa sambara, Moore, P. Z. S., 1865, p. 781.

" Male and female. Upperside dark maroon-brown; forewing with a series of seven semi-transparent white spots, three being small and obliquely subapical, the rest transverse to near posterior margin, and there joined by a short white longitudinal streak; hindwing with a broad central transverse whitish band bordered by a semi-circular discal series of black spots. Abdomen with whitish band. Cilia spotted with white. Underside as above but paler; the white band on the hindwing less defined but of a purer white. Palpi above and at the side, and legs in front blackish. Palpi and thorax beneath, legs and abdomen whitish; tip of the last brown.

Expanse: $1\frac{3}{4}$ inches.
Habitat: Darjeeling." (*Moore, P. Z. S., l. c.*)

This also occurs in Kumaon (*Doherty*), and in Sikkim (*de Nicéville; Elwes*).
In collections Indian Museum and de Nicéville.

131.—SATARUPA NARADA, MOORE.

Satarupa narada, Moore, J. A. S. B., 1884, p. 51.

" Upperside purpurascent violet-brown: forewing with three small upper and two lower subapical semi-diaphanous white spots, a small erect oval spot at lower end of the cell, a slightly larger quadrate spot on the disc between upper and middle medians and a broad band formed of three quadrate spots increasing in width from end of cell to posterior margin: hindwing with a broad white transverse medial band, the outer border with an ill-defined upper spot. Cilia edged with white. Underside marked as above; the hindwing with the band showing a more defined outer border and a well separated upper spot.

Expanse: $1\frac{4}{10}$ inches.
Habitat: Darjeeling, Sikkim.
Nearest allied to *S. bhagava*, but quite distinct." (*Moore, l. c.*)

Also recorded from Sikkim by Mr. Elwes who states it is a rare species; he also adds that the abdomen is white, excepting the anal extremity, and more slender than in other species of the genus.

Recorded by Mr. Doherty from Margherita, Assam.
In collections Indian Museum and de Nicéville.

132.—SATARUPA PHISARA, MOORE.

Satarupa phisara, Moore, J. A. S. B., 1884, p. 50.
Satarupa bhagava? de Nicéville, J. A. S. B., vol. lii, pt. 2, 1883, p. 90, n. 39, pl. x, fig. 14, ♀.

Satarupa phisara, Wood-Mason and de Nicéville, J. A. S. B., vol. lv, pt. 2, 1886, p. 390, n. 240, pl. xvii, fig. 4, ♂.

"Male. Upperside dark vinous-brown : forewing with two sometimes three or four, minute semi-diaphanous yellowish-white subapical spots, a small spot at lower end of the cell, a large quadrate spot below end of cell, and a small spot also between the base of upper and middle medians : a very indistinct greyish brown-speckled lunular fascia and a similar short fascia below the quadrate discal spot ; hindwing with a transverse subbasal pale yellowish band, and a curved submarginal indistinct greyish brown-speckled lunular fascia, which gives the discal area a macular appearance. Female : forewing marked as in male, the short fascia below the discal spot more distinct : hindwing with the transverse band somewhat broader, the discal area between it and the submarginal lunular fascia more distinctly macular, being transversed by pale veins. Underside as above, the markings more prominent. Abdomen with slender white narrow bands ; front of head and base of palpi and pectus orange-yellow, tip of palpi black.

Expanse : ♂ $1\frac{6}{10}$, ♀ $1\frac{6}{10}$ inches.
Habitat : Khasia Hills.
Allied to *S. bhagava* and *S. sambara*." (*Moore, l. c.*)

Also recorded from Sikkim (*de Nicéville*; *Elwes*), and Irangmara, Cachar (*Wood-Mason and de Nicéville*) ; occurs Yaw District, U. Burma.

In collections Indian Museum and de Nicéville.

133.—SATARUPA GOPALA, *MOORE.*

Goniloba gopala, Moore, Cat. Lep. Mus. E. I. C., i, p. 246.
Satarupa gopala, Moore, P. Z. S., 1865, p. 780, pl. xlii, fig. 1.

"Male and female dark maroon-brown. Upperside, forewing with a series of eight semi-transparent irregular-shaped whitish spots recurving transversely from costal margin before the apex to near posterior margin, being there joined by a white longitudinal streak ; a similar triangular shaped spot within the discoidal cell : hindwing pure white with the base and narrowly along anterior margin maroon-brown ; a double row of black marginal spots, the interspace between the rows being suffused with bluish-grey. Abdomen with broad white band ; tip brown. Cilia of hindwing white. Underside as above but with base of hindwing greyish-white and the double row of marginal spots more defined and blacker. Palpi, above brown, beneath orange-yellow. Thorax and legs beneath dull white ; legs in front black.

Expanse : ♂ $2\frac{1}{2}$, ♀ 3 inches.
Habitat : Darjeeling." (*Moore, l. c.*)

Mr. de Nicéville and Mr. Elwes also record this species from Sikkim.

In collections Indian Museum and de Nicéville.

GENUS XXX.—CALLIANA.

Calliana, Moore, P. Z. S., 1878, p. 686.

"Wings ample, broad. Forewing trigonal; cell long, broad; costal vein extending to one-third from apex; subcostal vein curved at end of the cell, first, second, third and fourth branches arising at equal distances before end of cell, fifth from its end, the three former terminating before the apex, fourth at the apex, and fifth below it; disco-cellulars angled close to the upper end; upper radial from the angle, lower from their middle; median vein three-branched, widely apart, the upper from lower end of cell, middle and lower branches at equal distances, the lower from near base of cell; submedian slightly recurved. Hindwing broadly ovate; subcostal two-branched, second before end of cell; disco-cellulars slightly angled; radial from their angle; median three-branched, middle branch from immediately before end of cell. Body small, robust, abdomen short. Palpi short, thick, pilose, ascending, third joint short, conical. Antennæ slender. Legs pilose." *(Moore, l. c.)*

134.—CALLIANA PIERIDOIDES, *MOORE.*

Calliana pieridoides, Moore, P. Z. S., 1878, p. 687, pl. xlv, fig. 2.

"Male. Upperside creamy-white, glossy; forewing with the apex broadly vinous-brown, the end of the median veins also with a vinous-brown-speckled spot: hindwing with a curved upper discal decreasing series of five purple-black spots, the upper one large and situated between the costal and subcostal veins; the end of the veins also with a small vinous-brown-speckled spot. Underside white; forewing with the costal base, a patch beyond end of the cell and the outer border pale vinous-brown: hindwing with a short black narrow streak at end of costal border, four large subasal and a curved discal series of seven purple-black spots, a marginal series of broad vinous-brown spots one at end of each vein. Thorax and abdomen white; collar, top of head palpi and tip of abdomen pale ferruginous; tip of palpi and antennæ black; legs pale ferruginous above, purple-brown beneath.

Expanse: 2⅝ inches.
Habitat:? N. E. Bengal." *(Moore, l. c.)* Buxa, Bhotan. de N.

Mr. Doherty, records several males from Margherita, Assam, but obtained no females, but states he once saw a male circling round a dark Hesperid which he failed to obtain and which possibly was the female.

GENUS XXXI.—TAGIADES.

Tagiades, Hübner, Verz. bek. Schmett., p. 108 (1816).
Pterygospidea, Wallengren, Rhop. Caffr., p. 53 (1857).
Tagiades, Moore, Lep. Cey., vol. 1, p. 175 (1880-81).
Tagiades, Distant, Rhop. Mal., p. 387 (1886).

"Forewing triangular; apex pointed; exterior margin oblique, very slightly convex, even; first, second and third subcostals at equal distances, first at one-third before end of the cell; disco-cellulars recurved inwardly, upper radial from angle near subcostal, lower from their middle; cell narrow, extending two-thirds the wing; middle median at one-sixth, lower at four-sixths before end of the cell, lower much curved at base within the cell; submedian slightly recurved: hindwing short, broadly produced hindward, exterior margin slightly scalloped towards anal angle; second subcostal at nearly one-third before end of the cell; disco-cellulars of equal length, nearly erect, radial from their middle; cell broad; middle median close to end and lower beyond one-third before end of the cell; submedian straight, internal recurved. Body short, thorax stout; palpi very compactly flattened laterally, terminal joint short, pointed; legs almost naked; antennæ slender, the club and lengthened tip very slender.

Type, *T. japetus.*" (*Moore, l. c.*)

135.—TAGIADES RAVI, *MOORE.*

Goniloba ravi, Moore, Cat. Lep. Mus. E. I. C., i, p. 246.
Pterygospidea ravi, Moore, P. Z. S., 1865, p. 779.
Tagiades ravi, Distant, Rhop. Mal., p. 388, pl. 34, fig. 1 ♂ (1886).
" Male and female fuliginous-brown.

Male. Upperside, forewing with three minute semi-transparent spots before the apex, two larger similar spots on the disc, diverging inward below the extremity of the cell; across the disc are three ill-defined blackish spots, and one before it near the base; apex and exterior margin blackish; hindwing with a curved series of small blackish discal spots. Underside brown: forewing with the semi-transparent spots as above; hindwing suffused with greyish-white, and having a curved series of small blackish discal spots. Palpi, body and legs beneath greyish-white. Female similar, but having two very minute additional spots beneath the subapical series, those on the disc being large and above the latter is a transverse spot at the extremity of the cell. Cilia brown throughout.

Expanse: ♂ 1⅜, ♀ 2 inches.
Habitat: Bengal," (*Moore, P. Z. S., l. c.*)

Also occurs in the Andamans and Nicobars (*Wood-Mason and de Nicéville*); Calcutta (*de Nicéville*); Tavoy (*Elwes and de Nicéville*); Cachar (*Wood-Mason and de Nicéville*); Orissa (*Taylor*).

" This species varies in the colour of the under-surface of the posterior wings, which in some specimens is wholly pale fuliginous-brown and in others more or less olivaceus-brown." (*Distant, l. c.*)

I have this species commonly from Rangoon and from Poungadaw, U. Burma. The apical spots on forewing vary from two to four.
In collections Indian Museum and de Nicéville.

135.—TAGIADES KHASIANA, *MOORE.*

Tagiades khasiana, Moore, J. A. S. B., 1884, p. 51.
" Male. Nearest to *T. ravi.* Of larger size, forewing comparatively more pointed at the apex: hinding also broader, and with more angular

apex. Upperside of a pale olivaceous-brown, the dusky margins less distinct on both wings, the apical and discal spots smaller. Female, upperside also paler than in *T. ravi*, the apical spots on forewing somewhat smaller, the cell spots similar, the two discal spots somewhat larger. Underside: forewing with the spots as above: hindwing more intensely whitish-grey, the discal black spots much smaller and less defined.

Expanse: ♂ 2, ♀ 2¼ inches.
Habitat: Khasia Hills; Shillong; Assam." (*Moore, l. c.*)
Also recorded from Calcutta (*de Nicéville*).
In collection Indian Museum.

137.—TAGIADES DISTANS, *MOORE*.

Tagiades distans, Moore, Lep. Cey., vol. i, p. 175, pl. 68, figs. 1, 1*a* (1880-81).

"Male and female. Upperside vinous-brown, discal areas suffused with greyish-brown: forewing with three minute semi-transparent white spots before the apex, two nearly obsolete below them, and two slightly larger spots in the disc, each bordered with black; a slight black spot below the cell near the base: hindwing with a greyish-white border extending from above anal angle to two-thirds the exterior margin; a blackish spot at end of the cell, and a curved series across the disc; cilia bordering the band also white.

Expanse: ♂ $1\frac{1}{10}$, ♀ $1\frac{9}{10}$ inches.

Allied to *T. ravi*, from which it may be distinguished by having the white band on the hindwing. The Javan *T obscurus*, Mabille, is also nearly allied, but differs in the absence of the discal semi-transparent spots on the forewing." (*Moore, l. c.*)

Occurs in Ceylon (*Wade, Hutchison*).
In collections Indian Museum and de Nicéville.

138.—TAGIADES OBSCURUS, *MABILLE*.

T. obscurus, Mabille, Ann. Soc. Ent. de France, fifth series, vol. vi, p. 274, n. 22 (1876).

"Forewings dusky, with three minute apical dots arranged somewhat in a circle on a black border. Hindwings with the base and exterior margin concolorous, but from the anal angle to the middle of the margin bluish-ashy, with three black dots on the border at the exterior margin. Underside hindwings broadly bluish-white with marginal black spots and the three black dots spoken of above but smaller.

This species belongs to that section in which the lower wings are not white with black spots, but brown and washed with bluish-white scales.

We have seen three males; there are very like the female of *T. gana*, Moore.

Habitat: Malay Archipelago, Java?" (*Mabille, l. c.*)
Expanse (not given.)

Also recorded from Cachar (*Wood-Mason and de Nicéville*); Nilgiris (*Hampson*); Orissa (*Taylor*).

I have this species commonly from the Kadur District, Mysore. The expanse of my specimens is 1·85 inches.

In collection Indian Museum.

139.—TAGIADES MEETANA, *MOORE*.

Tagiades meetana, Moore, P. Z. S., 1878, p. 842, pl. liii, fig. 1.

" Male. Allied to *T. obscurus*, Mabille, from Java: forewing above similar: hindwing differs in the marginal area of the anal angle being only slightly white-speckled between the veins; the cilia below it dark brown with a very narrow whitish inner line; on the underside there is less white, and the black border more lengthened and prominent, the cilia also being brown.

Expanse: $1\frac{2}{8}$ inches.

Habitat: Meetan, 3,000 ft. (U. Tenasserim) April." (*Moore, l. c.*)

Also recorded from Tavoy (*Elwes and de Nicéville*); Belgaum (*Swinhoe*).

I have several specimens of this species from Beeling, U. Tenasserim.

140.—TAGIADES ALICA, *MOORE*.

Tagiades alica, Moore, P. Z. S., 1877, p. 593, pl. lviii, fig. ii ♂.

Tagiades alica, Wood-Mason and de Nicéville, J. A. S. B., vol. xlix, pt. 2, 1880, p. 241, n. 79, ♀.

" Male. Allied to *T. obscurus*, Mabille, from Java, but of a blacker colour, the forewing being more pointed, the hindwing more produced at the anal angle. Markings above similar; on the underside the hindwing has more prominent black discal marks, and a much broader black border on the grey portion of the outer margin.

Expanse: $1\frac{7}{8}$ inches.

Habitat: S. Andamans." (*Moore, l. c.*)

" Female. Above lighter, the dark markings consequently appearing more prominent.

The anterior wing has a minute transparent speck behind the three sub-apical ones, a very indistinct and small double whitish spot near the end of the cell on the upperside, and two discal whitish spots on the underside, the anterior one of which only is partially transparent and visible on the upperside. The posterior wing is less white above and has the anal angle rounded as in *T. obscurus*." (*Wood-Mason and de Nicéville, l. c.*)

Also recorded from Andamans by Messrs. Wood-Mason and de Nicéville, and from Tavoy by Messrs. Elwes and de Nicéville, who note that this species " Differs from *T. meetana* in having the upperside of the hindwing broadly white, not slightly white-speckled between the veins, as in that species, and the cilia white throughout instead of being tipped with dark brown."

In collections Indian Museum and de Nicéville.

141.—TAGIADES MENAKA, *MOORE*.

Goniloba menaka, Moore, Cat. Lep. Mus. E. I. C., p. 246.
Pterygosidea menaka, Moore, P. Z. S., 1865, p. 778.

"Male and female dark brown.

Male. Forewing with six minute semi-transparent white spots, recurving before the apex, and two similar spots from near middle of the costa: hindwing with a large discal quadrate space pure white, a series of brown marginal spots with their interspaces greyish, and within the white space too smaller paler spots. Base of abdomen white.

Female. Marked as in the male, but having the semi-transparent spots on the forewing larger, the black marginal spots of hindwing more distinct and apart, and with the two spots on the white space. Underside as the upperside both sexes having the base of the hindwings, palpi beneath, legs and body greyish-white, and the black spots on the hindwing extending towards the base of the anterior margin. Palpi above brown.

Expanse: ♂ 1½, ♀ 1¾ inches.
Habitat: N. E. Bengal." (*Moore, P. Z. S., l. c.*)

Also occurs in Sikkim (*de Nicéville*); Andamans (*Wood-Mason and de Nicéville*); Kangra (*Moore*).

In collection Indian Museum.

142.—TAGIADES ATTICUS, *FABRICIUS*.

Hespeiria atticus, Fabricius, Ent. Syst., vol. iii, pt. i, p. 339.
Tagiades atticus, Moore, Lep. Cey., vol. i, p. 175, pl. 68, fig. 2 (1880-81).
Tagiades atticus, var. *calligana*, Distant, Rhop. Mal., p. 387, pl. xxxiv, fig. 6 (1886).

"Upperside dark vinous-brown; forewing with seven minute semi-transparent white spots recurving before the apex, two within end of the cell and one above its end; hindwing with a large pure white quadrate anal patch bordered before the apex by two rather large black spots, and along exterior margin by four smaller spots; cilia bordering the patch also white. Underside paler, the white patch extending to base of hindwing. Palpi legs and body beneath greyish-white.

Expanse: ♂ 1½, ♀ 1¾ inches." (*Moore, l. c.*)

Occurs in Ceylon (*Hutchison, Wade, Macwood*); Orissa (*Taylor*); Cachar (*Wood-Mason and de Nicéville*); Sikkim (*de Nicéville, Elwes*); Nilgiris (*Hampson*).

Doherty records this species from Kumaon and considers it synonymous with *menaka*, Moore; Mr. Elwes also considers it synonymous with *menaka*, and states that specimens from Bhutan have no black spots within the white patch of the hindwing. Mr. de Nicéville states that the only difference he can detect between the two species is that *T. atticus* has two spots in the cell of the forewing, and *menaka* only one, and that though both occur in Sikkim he has never received both species from any other locality.

On the whole the probability seems to be that this and the preceding are identical, in which view Mr. de Nicéville now agrees.

In collection Indian Museum and de Nicéville.

143.—TAGIADES GANA, *MOORE.*

Pterygospidea gana, Moore, P. Z. S., 1865, p. 180.

Tagiades gana, Distant, Rhop. Mal., p. 388, pl. xxxiv, fig. 2, ♂ (1886).

" Male and female dark brown.

Male. Upperside with three minute semi-transparent spots obliquely before the apex ; a transverse discal series of streaks, a small patch within the cell, one near the base of the wing, and exterior margin blackish ; hindwing with the lower third pure white which is straightly separated from the brown of the basal portion ; apical margin and three spots on the upper part of the disc, black, and two spots on the middle of the white anterior margin. Underside paler brown, semi-transparent spots on forewing as above : hindwing white suffused with brown along the anterior margin : upper discal and marginal spots as above black.

Female paler. Upperside somewhat greyish-brown ; forewing with spots and blackish discal streaks, and hindwing with upper discal spots as in male : exterior margin of hindwing greyish-white. Underside as in male. Cilia of both sexes pure white on the lower portion of the hindwing, the rest brown.

Expanse : ♂ $1\frac{3}{4}$, ♀ $1\frac{7}{8}$ inches.

Habitat : Bengal." (*Moore, l. c.*)

Also occurs in Sikkim (*de Nicéville* ; *Elwes*).

In collections Indian Museum and de Nicéville.

144.—TAGIADES MINUTA, *MOORE.*

Tagiades minuta, Moore, Annals and Magazine Nat. Hist., series iv, vol. xx, p. 343 (1877).

Tagiades minuta, Moore, Lep. Cey., vol. i, p. 176, pl. 68, fig. 44*a*, (1880-81).

" Male and female. Upperside, dark olive-brown. Cilia of hindwing white, of forewing slightly whitish at posterior angle. Underside, forewing greyish-white on middle of hind margin ; hindwing greyish-white, except along anterior border ; a small blackish spot at end of the cell, and a medial discal series of spots ; outer margin brown speckled. Palpi body and legs beneath grey-white. Legs above brown.

Expanse : ♂ $1\frac{1}{2}$, ♀ $1\frac{3}{3}$ inches." (*Moore, l. c. in Lep. Cey.*) Occurs in Ceylon (*Wade*).

In collections Indian Museum and de Nicéville.

145.—TAGIADES PRALAYA, *MOORE*.

Goniloba pralaya, Moore, Cat. Lep. E. I. C., vol. i, p. 246 (1857).
Pterygospidea pralaya, Moore, P. Z. S., 1865, p. 779.

" Male and female yellowish-brown, veins paler on the disc. Upperside: forewing with numerous series of variously-shaped small semitransparent white spots, five of which are placed obliquely before the apex, the rest disposed from near middle of the costa and extending across the disc, three of which are in the form of elongated streaks: hindwing with the exterior half orange-yellow, having a subbasal series of longitudinal black streaks between the yellow veins: apex with suffused blackish spots. Abdomen with yellow segmental bands. Underside as above: forewing with the veins from the base lined with yellow; a series of submarginal ill-defined yellow spots. Palpi, body, and legs yellowish-brown.

Expanse: ♂ 1¾; ♀ 2 inches.
Habitat: Bengal." (*Moore, P. Z. S., l. c.*)

Also recorded from Ahsown, Upper Tenasserim (*Limborg*); Sikkim (*Elwes*).

T. trichoneura of Felder is distinguished by the pearly-white colour of the hindwing beneath.

In collections Indian Museum and de Nicéville.

146.—TAGIADES TRICHONEURA, *FELDER*.

Pterygospidea trichoneura, Felder, Wien. Ent. Mon., vol. iv, p. 402, n. 31 (1860); id., Reise Nov. Lep., vol. iii, pl. 73, figs. 14, 15 (1867).
Tagiades trichoneura, Kheil, Rhop. der Insel Nias, p. 38, n. 149 (1881).
Tagiades trichoneura, (*var.*), Distant, Rhop. Mal., p. 389, pl. xxxiv, fig. 20 (1886).

" Wings above dark purplish-brown; anterior wings with the neuration more or less greyish, and with thirteen pale greyish spots, situated two in and two above cell, four in oblique series beneath cell, and five subapical (there is sometimes a small fourteenth subcostal spot as in the specimen figured). Posterior wings with the outer margin from the upper median nervule to anal angle broadly orange-yellow, and the same colour is extended narrowly for a short distance along the margin as far as the lower subcostal nervule, two elongate spots in cell, a discal series of elongate spots and the apical margin dark purplish. Anterior wings beneath as above, but with a few additional greyish spots; posterior wings beneath pearly-white, the costal and apical areas purplish-brown, containing a series of dark purplish elongate spots. Body above purplish-brown, the abdominal segments margined with greyish; body beneath with legs greyish.

Expanse: 33 to 44 millim.
Habitat: Malay Peninsula, Perak (*Künstler, Calc. Mus.*); Malacca (*coll. Staudinger* and *Felder*); Nias Island (*Kheil*); Java (*Felder*)." (*Distant, l.c.*)

I have two specimens of this species from Beeling, Upper Tenasserim, they are absolutely identical with Distant's figure, and have the underside of the hindwing pearly-white and not ochraceous as in *T. pralaya*.

There are single specimens of this and the preceding species in the Phayre Museum, Rangoon, the former from the Karen Hills and the latter from Tavoy. They are readily separable by the colour of the underside of the hindwing.

147.—TAGIADES HELFERI, *FELDER*.

Pterygospidea helferi, Felder, Verh. zool.-bot. Gesellsch. Wien., vol. xii, p. 483, n. 116 (1862).

"Wings above fuscous, forewing on both sides with minute spots beyond the disc and some submarginal hyaline dots, obsolete discal spots dusky-black, hindwing on both sides with a discoidal spot and others exterior more or less obsolete dusky-black, underside with the inner margin washed with bluish-white, hind margin obsoletely fuscous. ♂.

Allied to *P. japetus*, Cramer. Occurs with slight variations on the Indian continent. The Nicobar specimens are smaller than the continental ones. The species has a certain resemblance to some species of *Eudamus*.

Habitat: Pulo Milu [Nicobars]." (*Felder, l. c.*)

Mr. de Nicéville informs me that this species occurs commonly in the Nicobar group of islands, and has been obtained on Kamorta, Nankowri, Katschall, Kondul, and Meroe, by the late Mr. F. A. de Roepstorff and Mr. Man.

In collections Indian Museum and de Nicéville.

GENUS XXXII.—ABARATHA.

Abaratha, Moore, Lep. Cey., vol. i, p. 181 (1881).

Abaratha, Distant, Rhop. Mal., p. 391 (1886).

"Forewing triangular; apex pointed; exterior margin short, oblique, slightly convex and uneven; posterior margin short; first and fourth subcostals at equal distances apart, first emitted at one-third before end of the cell; disco-cellulars inwardly oblique, upper radial from angle near subcostal, lower from their middle; cell two-thirds the wing; middle median at one-fourth and lower at two-thirds before end of the cell; submedian much recurved: hindwing short, apex angular, exterior margin sinuously angular below the apex and in the middle; abdominal margin rather long; second subcostal at one-fourth before end of the cell; discocellulars nearly erect, upper shortest, radial from their middle; cell broad, middle median from close to end of the cell and lower at beyond one-third before the end; submedian straight, internal recurved. Body short, robust; palpi laterally compressed in front, terminal joint short, conical; legs naked; antennæ hooked, club thick and bluntly pointed.

Type: *A. ransonnetii*." (*Moore, l. c.*)

148.—ABARATHA RANSONNETII, *FELDER*.

Pterygospidea ransonnetii, Felder, Verh. zool.-bot. Gesellsch., 1868, p. 284.

Pterygospidea potiphera, Hewitson, Exot. Butt., vol. v, *Pterygospidea*, pl. 1, fig. 7 (1873).

Abaratha ransonnetii, Moore, Lep. Cey., vol. 1, p. 182, pl. 97, fig. 1 (1881).

Abaratha taylorii, de Nicéville, J. A. S. B., vol. lii, pt. 2, p. 88, pl. x, fig. 13, *male* (1883).

" Upperside fuliginous ochreous-brown. Male; forewing with three small semi-transparent white spots before the apex (and sometimes one or two very minute spots obliquely below them), two spots within end of the cell, a slender spot between the upper and middle median veins, a larger spot between the latter vein and submedian, and followed below it by two small obliquely disposed spots; a marginal double row of pale indistinct small lunules; hindwing with a broad medial discal macular pale ochreous band traversed by brown veins and a spot within end of the cell, the outer discal area suffused with grey-brown. Cilia alternated with white. Female; forewing with the spots and marginal lunules, and the macular band on hindwing more prominent, the latter also more distinctly bordered with grey. Underside: forewing paler brown; the basal area greyish-white, the spots with clouded black outer borders; hindwing greyish-white, the outer margin only being brown, traversed by a curved discal series of small blackish spots.

Expanse: ♂ $1\frac{3}{8}$; ♀ $1\frac{5}{8}$ inches." (*Moore, l. c.*)

Occurs Ceylon (*Hutchison, Wade, Mackwood*); Orissa, (*Taylor*); Nilgiris, (*Hampson*).

The dry-season form which has been named *A. taylorii* by de Nicéville differs in being ochreous not dark brown above, and in having the disc of the hindwing unmarked with a group of ochreous spots and streaks.

A similar variation has been noted by Mr. de Nicéville in *C. tissa*, a not very distantly allied species, and in both cases it is the dry-season form which is the paler.

In collections Indian Museum and de Nicéville.

149.—ABARATHA SARAYA, *DOHERTY*.

Abaratha saraya, Doherty, J. A. S. B., 1886, p. 138.

" Agrees with Mr. Moore's description of the genus *Abaratha* except that the apex of the forewing is not acute but right-angled, that of the hindwing decidedly rounded. The outer margin of the hindwing is also more scalloped, and less irregularly angulate. Above fuscous, with the following tawny-ochreous marks :—a line of streaks just within the margin; a line of square spots from the costa to the upper median branch, continued to the hind-margin by a series of large and more irregular blurs, removed further from the outer margin; a dull area just beyond the cell, from costa to the middle median; three large irregular spots occupying the middle of the cell, and the two spaces below between the submedian and the middle median. Also the following translucent spots:—five apical ones, the upper three elongate and approximate; one at the end of the cell almost bifid, with a dot on the costa above it; four on the disc from the submedian to the upper median, the second from above largest of all and adjoining that at the end of the cell. Also one in the cell near the base. All these are surrounded by blackish rings above and below. A blackish marginal line; cilia long, alternately black and whitish. Hindwing rusty-ochreous, with a marginal dark line, and a discal, a cellular, and a submarginal row of dusky spots but no translucent ones. Below paler ochreous, without any rufous tinge, the translucent spots set in small blackish patches; a submarginal line of joined dusky spots, and a dark streak near the base from the submedian to the median veins. Hindwing with a black transverse streak at the end of the cell, a fainter one near the base of the cell, and a circle of large and conspicuous black spots nine in all round the disc, whereof two are between the costal and the subcostal and two between the median and the submedian veins; whitish hairs at the extreme base. Body dull ferruginous above, whitish below. Female unknown.

Differs from *A. ransonnetii*, Felder, its nearest ally, in the absence of all white on the disc below. In colouring it is somewhat intermediate between *A. ransonnetii* and the curious *A. agama* of Sikkim, which seems to mimic *Argynnis isaea*.

One male, Bágheswar on the Sarju, 3,500 ft." [Kumaon]. (*Doherty, l. c.*)

150.— ABARATHA SYRICHTHUS, *FELDER*.

Pterygospidea syrichthus, Felder, Reise Novara, Lep., vol. iii, p. 530, n. 938, pl. lxxii, figs. 22, 23 (1867).

Pyrgus agama, Moore, MS., Cat. Lep. Mus. E. I. C., vol. 1, p. 249, n. 556, pl. vii, figs. 1, *larva*; 1a, *pupa* (1857).

The Latin in which this species is described is so very obscure that I have thought it better to quote it in full instead of giving a translation of it, as without examples of the species it is very difficult to translate it correctly.

" ♂. Alæ ciliis latiusculis, fuscis, albo intercisis, supra omnes fuscæ, anticæ maculis duabus cellulæ, totidem interioribus basalibus, lunula disco-cellulari, maculis submarginalibus in serie fracta aliisque antemarginalibus seriatis minoribus albidis, macula cellulari excisa, quatuor interioribus (secunda inter ramum medianum primum et secundum majore) tribusque congestis subapicalibus subhyalinis, in certo situ argenteo micantibus, posticæ macula cellulari, duabus anticis, aliis inaequalibus proxime pone cellulam, dein septem submarginalibus aliisque antemarginalibus in serie margini parallela albidis.

Alæ anticæ subtus pallidiores, maculis iisdem sed majoribus, basi glaucescente, posticæ albæ, ad basin glaucescentes, margine antico fusco atomato, externo excise fusco, maculis duabus anticis, virgula sæpe evanescente maculaque cellaribus, maculis sex exterioribus aliisque grossioribus antemarginalibus fuscis.

Habitat: Java." (*Felder, l. c.*)

Expanse not given [1·7″ Felder's figure].

Mr. Elwes (Trans. Ent. Soc. Lond., 1888, p. 458), says, " Mr. Doherty mentions this species as having been taken in Sikkim. Möller also has taken what I believe to be this species in the Terai during the rains."

Mr. Hampson [J. A. S. B., 1888, p. 368], states that a single specimen was taken by Mr. Alfred Lindsay on the southern slopes of the Nilgiri Hills at 3,000 feet elevation.

Since writing the above I have, during May, obtained several specimens of this species at Loungat at the foot of the Chin Hills in Upper Burma. They settled invariably on the ground with their wings extended flat. The following description is taken from my specimens:—

Expanse: 1·5 inches.

Upperside, brown with numerous white and straw-coloured spots. *Forewing* with a minute white spot near base of cell, two conjoined straw-coloured spots beyond it also minute, a large square outwardly-indented white spot near end of cell, with a minute linear straw-coloured one above it, and a lunate straw-coloured spot closing the cell; a round white spot beneath this last one, with three rounded white spots—of which the centre one is minute—nearer base of wing and almost directly below the large white spot in cell, also three small straw-coloured spots still nearer base of wing; a discal row of nine spots, curving across the wing, beyond the cell, of which, the upper three are white, conjoined, and linear, the next two also white and minute, and the remainder straw-coloured; four apical submarginal straw-coloured spots, and a complete row of eight similar marginal ones. *Hindwing*, with a spot near base, another lunate one in cell, and two complete marginal rows, all straw-coloured; inner margin heavily clothed with long fuscous setæ. Underside, *forewing* densely clothed at base with greyish-white scales, the spots as above but all white, the yellow ones of the upperside being opaque and the white ones semi-transparent. *Hindwing*, white, with the spots as above but more numerous and black, a prominent anteciliary black line. *Cilia* above and below alternated with black and white. *Antennæ* fuscous; *body* fuscous above and whitish beneath.

In collections Indian Museum and de Nicéville.

GENUS XXXIII.—CTENOPTILUM.

Ctenoptilum, de Nicéville, Journ. Bomb. Nat. Hist. Soc., vol. v, p. 220 (1890).

" Forewing, narrow, elongated; *costa* arched at base, then straight to apex; *apex* acute; *outer margin* at right angles to costa from apex to termination of third median nervule, this portion also being slightly excavated, from third median nervule to inner angle strongly inwardly oblique, also slightly concave; *inner angle* rather acute; *inner margin* sinuous; *costal nervure* very short, not nearly reaching opposite to the apex of the discoidal cell; *first, second,* and *third subcostal nervules* also very short, rapidly reaching the costa, *fourth* subcostal long, extending to apex of wing, the bases of all the subcostals nearly equi-distant; terminal portion of *subcostal nervure* reaching outer margin below apex of wing; *discoidal cell* long, narrow, reaching to more than half though less than two-thirds the length of the wing; *upper disco-cellular nervule* short, straight, outwardly oblique; *middle* and *lower* disco-cellulars straight, slightly inwardly oblique, the lower a little longer than the middle; *second median* nervule arising considerably before the lower end of the cell; first median arising much nearer to the base of the wing than to the lower end of the cell; *submedian nervure* sinuous, following the outline of the inner margin; *internal* nervure short, running into the submedian nervure as usual. Hindwing, with the base of the *costa* much produced, thence gently curving to apex; *outer margin* slightly produced tooth-like at apex of first subcostal nervule, very strongly at third median nervule, thence inwardly oblique to anal angle and slightly concave; *anal angle* rounded; *inner margin* nearly straight; *costal nervure* curved, reaching the apex of the wing; *first subcostal nervule* originating long before the apex of the discoidal cell; *disco-cellular* nervules of equal length, almost straight, slightly outwardly oblique; *discoidal* nervule fine but quite distinct; *second median* nervule given off close to the lower end of the cell; *first* median arising nearer to lower end of the cell than to the base of the wing; *submedian* and *internal nervures* almost straight. *Antennæ* about half as long as the costa of the forewing, with a well-formed club; *thorax* rather stout; *abdomen* rather slender, not quite reaching to anal angle of hindwing. Male with no secondary sexual characters on the wings, but with a dense tuft of hairs attached to the anterior end of the tibia of the hind legs, the hairs extending to the apex of the first joint of the tarsus. Female like the male, except that the wings are rather larger and broader. Type, the "*Achlyodes*" *vasava* of Moore.

Ctenoptilum is evidently nearest allied to *Odontoptilum*, mihi, but differs considerably in the outline of the wings, in the shortness of the costal nervure and subcostal nervules of the forewing, and especially in the setose clothing of the legs of the male, in *Ctenoptilum* the hairs are much longer and do not form such a large and dense mass as in *Odontoptilum*, and are attached to the tibia of the hindlegs instead of to the coxa of the forelegs.

As far as is known to me, the genus contains but two species, which occur in Sikkim, Assam, and Burma. They probably rest with wide outspread wings." (*de Nicéville, l. c.*)

151.—CTENOPTILUM VASAVA, *MOORE.*

Achlyodes vasava, Moore, P. Z. S., 1865, p. 786.
Ctenoptilum vasava, de Nicéville, Journ. Bomb. Nat. Hist. Soc., vol. v, p. 221 (1890).

"Upperside dull ferruginous, palest on the hindwing: forewing slightly suffused with blackish along the posterior margin; an irregular series of variously-shaped semi-transparent spots disposed across the disc, with an exterior blackish transverse streak: hindwing with the base suffused with blackish, a subbasal agglomerated series of irregularly-shaped semi-transparent spots. Underside paler, marked as above, but without the transverse black outer streak on the forewing.

Palpi and body beneath whitish. Legs ferruginous.
Expanse: 1½ inches.
Habitat: Darjeeling." (*Moore, l. c.*)
Also recorded from Upper Assam (*Butler*); Sikkim (*Elwes*).

According to Mr. Elwes a very nearly allied if not identical species occurs in N. China, and he states he has two specimens of another possibly new species (presumably allied to this and probably the species described next below), from Akyab and Tenasserim.

I obtained a single specimen from Beeling, Upper Tenasserim, identified by Mr. de Nicéville.

In collections Indian Museum and de Nicéville.

152.—CTENOPTILUM MULTIGUTTATA, *DE NICÉVILLE.*

Ctenoptilum multiguttata, de Nicéville, Journ. Bomb. Nat. Hist. Soc., vol. v, p. 221, n. 16, pl. E, fig. 10, ♂ (1890).

Habitat: Burma.
Expanse: ♂, ♀, 1·5 inches.

"Description: Male. Upperside, *both wings* reddish-ochreous, with numerous lustrous semi-transparent white spots. *Forewing* with a small spot in the discoidal cell towards the base, with an elongated spot below it in the submedian interspace, sometimes divided into two; a very large spot at the end of the discoidal cell, outwardly anteriorly deeply incised; two spots above it just below the costa divided by the first subcostal nervule; a discal curved series of ten spots, of which the first four are of nearly equal size (the uppermost the smallest), divided by the subcostal nervules, followed by a rapidly increasing series of four spots, one in each interspace; then by two spots, the lower three times as large as the upper, in the submedian interspace; a submarginal obscure fuscous band. *Hindwing* with a clump of spots shewing great diversity in shape and size occupying the middle of the wing; a submarginal dark fascia as in the forewing. Underside, *both wings* coloured and marked as above, but all except the outer margin thickly frosted with grey, which appearance is found on examination under a strong lens to be due to the presence of long white hair-like scales scattered somewhat sparsely over the surface. Female differs from the male only in being rather paler, and the submarginal dark fascia above more prominent.

Near to the "*Achlyodes*" *vasava* of Moore from Sikkim, Assam [and Upper Tenasserim], from which it differs in its more reddish less ochreous ground-colour, considerably less angulated wings, the forewing on the upperside on the disc and base not suffused with black, the third spot of the discal series equal in length to the spot on either side of it, instead of, as in *C. vasava*, being greatly lengthened out and ending in a point just below the costa; and all the spots on the hindwing smaller, especially those in the discoidal cell, the middle spot of *C. vasava* being divided into two in *C. multiguttata*.

Described from two male specimens from the Meplay Valley, taken in February, another from the Donat Range, taken in January, and a female from the valley of the Houngdarou taken in March, all in Upper Burma, by Majors C. T. Bingham and C. H. E. Adamson. This is probably one of the species referred to by Mr. H. J. Elwes as from Akyab and Tenasserim in Trans. Ent. Soc. Lond., 1888, p. 458, n. 512." (*de Nicéville, l. c.*)

GENUS XXXIV.—ODONTOPTILUM.

Odontoptilum, de Nicéville, Journ. Bomb. Nat. Hist. Soc., vol. v, p. 217 (1890).

"Forewing, elongated; *costa* gently arched; *apex* acute; *outer margin* at first at right angles to the costa, then directed strongly obliquely inwardly; *inner angle* rather acute, *inner margin* short, sinuous, of about the same length as the outer margin; *costal nervure* not reaching to opposite the apex of the discoidal cell; all four *subcostal nervules* with their bases about equally distant one from the other; the *fourth* subcostal reaching the apex of the wing; the terminal portion of the *subcostal nervure* ending on the outer margin considerably below the apex of the wing; *upper disco-cellular nervule* stout, straight, short, outwardly oblique; *middle* disco-cellular shorter than lower; *lower* disco-cellular straight, in the same straight line as the middle, both strongly inwardly oblique; *median* nervules arising very far apart, *second* median arising long before the lower end of the cell; *first* median arising one-third of the length of the cell from the base of the wing; *submedian nervure* sinuous; *internal* nervure short and running into the submedian nervure, both as usual. Hindwing, *costa* short, much produced at base, then straight to apex; *outer margin* strongly excavated between terminal points of costal nervure and first subcostal nervule, then arched to anal angle; *anal angle* slightly produced lobe-wise; *inner margin* sinuous; *costal nervure* gently arched, ending at apex of wing; *first subcostal nervule* arising some distance before the apex of the discoidal cell; *upper disco-cellular* nervule straight, slightly outwardly oblique, in the same straight line as the lower; *lower* disco-cellular longer than the upper; *discoidal* nervule fine but quite distinct; *second median* nervule given off just before the lower end of the cell; *first* median given off slightly nearer to the lower end of the cell than to the base of the wing; *submedian* and *internal nervures* straight. *Antennæ* less than half as long as the costa of the forewing, with a well-formed club and hooked tip; *thorax* stout, *abdomen* rather stout, not nearly as long as the abdominal margin of the hindwing. Sexes alike, *male* with no secondary sexual characters on the wings, but the fore legs are furnished with a very thick tuft of hairs attached to the anterior end of the coxa, the hairs being slightly longer than that joint. Type, "*Achlyodes*" *sura*, Moore.

Mr. Moore placed the type species of *Odontoptilum* in the genus *Achlyodes*, Hübner (1816), of which the type is the South American *fredericus*, Hübner, with which *O. sura* will probably be found to have but slight connection. Mr. Distant placed *O. sura* in the genus *Abaratha*, Moore (1881), of which the *Pterygospidea ransonnetii* of Felder is the type. In that genus the forelegs of the males are furnished with a tuft of long setæ which are also found in *O. sura*, but are very much shorter and much more dense in the latter. The type species of *Abaratha* and *Odontoptilum* differ also in the shape of the discoidal cell of both wings; in the former in the forewing the middle disco-cellular nervule is upright, and therefore forms an obtuse angle with the inwardly oblique lower disco-cellular; in the latter the two veins are in one straight line; in the hindwing of the former the lower disco-cellular is quite upright, thus forming an angle with the upper outwardly oblique disco-cellular; in the latter both are in one straight line, and are outwardly oblique. Otherwise there does not appear to be much difference between the two genera either in neuration or outline of the wings. Mr. Kirby places *O. sura* in the genus *Antigonus* of Hübner (1816), of which the *nearchus* of Latrielle from South America is the type. It is very improbable that this species either is congeneric with *sura*.

The genus *Odontoptilum* occurs all along the outer ranges of the Himalayas, in South India, in Assam, Burma, the Malay Peninsula, Sumatra, Borneo, Celebes, the Philippine Isles, and China. They rest with wide outspread wings." (*de Nicéville, l. c.*)

153.—ODONTOPTILUM SURA, *MOORE*.

Achlyodes sura, Moore, P. Z. S., 1865, p. 786.

Abaratha sura, Distant, Rhop. Mal., p. 391, pl. xxxiv, fig. 16, ♂ (1886).

Odontoptilum sura, de Nicéville, Journ. Bomb. Nat. Hist. Soc., vol. v, p. 218 (1890).

"Male and female vinaceous-brown, palest on the hindwing. Male, forewing dull chestnut-brown along exterior margin, with a black transverse band one-third from the base; a geminated semi-transparent spot on costa before the apex, surrounded by suffused black; a semi-transparent lunule and a small spot on the lower part of the disc, bordered without by a black band: hindwing with a transverse subbasal, an elbowed discal, and a lower submarginal purplish-white line; apex of wing with suffused black patch and lower marginal blackish pale-bordered spots. Underside brown, forewing suffused with greyish-white at the base; markings as above: hindwing greyish-white, the transverse lines less defined, marginal spots blacker, and a blackish spot near base of wing. Palpi above black. Palpi and body beneath and legs greyish-white. Female paler, marked on upper-and undersides as in male.

Expanse: ♂ $1\frac{8}{8}$; ♀ $1\frac{7}{8}$ inches.

Habitat: N.-E. Bengal." (*Moore, l.c.*)

Also recorded from Cachar (*Wood-Mason and de Nicéville*); Orissa (*Taylor*); Kumaon (*Doherty*); Tavoy (*Elwes and de Nicéville*); Nilgiris (*Hampson*); Sikkim (*de Nicéville*; *Elwes*).

I have this species commonly from Rangoon.

In collections Indian Museum and de Nicéville.

GENUS XXXV.—DARPA.

Darpa, Moore, P. Z. S., 1865, p. 781.

" Palpi stout, densely pilose; third joint small, conical, hidden by the hairs. Antennæ moderate, hooked at the tip; legs short; femora slightly pilose; fore tibiæ short and rather stout; mid tibiæ slightly and hind tibiæ densely pilose; mid and hind tibiæ armed with a pair of short spurs (the usual second pair on the latter invisible). Body stout, abdomen short. Wings small; costa of forewing nearly straight; exterior margin irregularly scalloped, produced in the middle: hindwing somewhat quadrate; exterior margin irregularly scalloped, produced to an angle in the middle." (*Moore, l. c.*)

154.—DARPA HANRIA, *MOORE.*

Darpa hanria, Moore, P. Z. S., 1865, p. 781, pl. xlii, fig. 2.

" Upperside black, with minute bluish-grey scales in patches between the veins and narrowly along the veins: forewing with a series of semi-transparent irregular-shaped spots, the largest of which is within the extremity of the cell, others above and beneath it; before the apex are three small conjugated spots: hindwing with a pale yellow space broadly occupying the lower portion of the exterior margin: apical and two medial angles with a black spot; base of wing adorned with very long brown and yellow hairs. Underside paler; forewing with markings as above, densely irrorated with bluish-white scales: hindwing bluish-white at the base, yellowish-white exteriorly; anterior margin and apex blackish; below the anterior margin are three black spots, and a spot on the two medial angles of exterior margin. Palpi and body beneath and legs white.

Expanse: $1\frac{3}{8}$ inches.

Habitat: N.-E. Bengal." (*Moore, l. c.*).
Also recorded from Sikkim (*Elwes*; *Möller*).
In collections Indian Museum and de Nicéville.

GENUS XXXVI.—ERIONOTA.

Erionota, Mabille, Ann. Soc. Ent. Belg., vol. xxi, p. 34 (1878).
Erionota, Distant, Rhop. Mal., p. 390 (1886).

"Anterior wings moderately long, the inner margin longer than the outer, costal margin very slightly sinuate, inner margin distinctly sinuate. Costal nervure terminating on costa nearly opposite end of cell; fifth subcostal nervule emitted at about end of cell: disco-cellular nervules moderately oblique, the upper and lower subequal in length; second median nervule with its base more than twice as far apart from that of lower as from that of upper median nervule. Posterior wings about as broad as long, the costal margin obliquely convex, the outer margin irregularly rounded and slightly sinuate towards anal angle. Subcostal nervules bifurcating at about half the distance before end of cell; second median nervule emitted at rather more than twice the distance from lower than from upper median nervule, which starts from end of cell. Body long, robust and pilose; palpi large, broad and considerably compressed, the terminal joint very short; antennæ slender, the apex moderately thickened and curved, not strongly hooked." (*Distant, l. c.*)

155.—ERIONOTA THRAX, *LINNÆUS*.

Papilio thrax, Linnæus, Syst. Nat., vol. i, pt. 2, p. 794, n. 260 (1767).
Telegonus thrax, de Nicéville, J. A. S. B., vol. li, pt. 2, p. 65, n. 196 (1882).
Erionota thrax, Distant, Rhop. Mal., p. 393, pl. xxxiv, fig. 17 (1886).

"Male and female. Wings above chocolate-brown; anterior wings with three discal pale ochraceous spots, situate one crossing cell, another beneath cell and between the two lower median nervules, and the third and smallest between the first and second median nervules; posterior wings with the fringe greyish-ochraceous. Wings beneath paler than above; posterior wings with a discal, rounded, macular, darker fascia. Body and legs more or less concolorous with wings.

Expanse ♂ and ♀ 63 to 78 millims." (*Distant, l. c.*)
Recorded from Calcutta (*de Nicéville*); Bengal (*Moore*); Sikkim (*Elwes*). In collections Indian Museum and de Nicéville.

156.—ERIONOTA ACROLEUCA, *WOOD-MASON AND DE NICÉVILLE.*

Telegonus acroleucus, Wood-Mason and de Nicéville, Proc. A. S. B., August, 1881, p. 143.
Hesperia hiraca, Moore, Trans. Ent. Soc. Lond., September, 1880, p. 313, ♀.
Hesperia acroleuca, Wood-Mason and de Nicéville, J. A. S. B., vol. l, pt. 2, p. 260, n. 126 (1881).

"Male. Wings above dark brown slightly suffused with vinous. Anterior wings tipped with ashy-white, and with three large semi-transparent pale yellow quadrangular lustrous spots arranged as in *E. thrax*, namely, one in the cell with its outer margin bifestooned, and its inner biscalloped, another larger and elongated below and partly under this between the first and second median veinlets, and a third, the smallest of the three, rhomboidal, between the second and third median veinlets, and with the cilia dusky at the apex, but gradually becoming pale yellow towards the inner angle. Posterior wings darker towards the outer margin, with all the cilia pale yellow. Wings below paler and duller, suffused with purple on the disc, and ornamented—especially on the medial area of the posterior pair—with scattered ochreous setiform scales. Antennæ black, with the straight portion of the club broadly and conspicuously incompletely ringed with cretaceous-white, and the much shorter terminal hooked portion red internally. Female, differs from the male only in its larger size, and in not having the anterior wings tipped with ashy, nor the antennæ nearly so conspicuously ringed with white. ♂ ♀ Eyes blood-red. Palpi with the terminal joint rudimentary.

Expanse: ♂ 2·26; ♀ 2·32 inches.

Habitat: Andamans." (*Wood-Mason and de Nicéville, J. A. S. B., l. c.*)

Also recorded from Sikkim (*de Nicéville*).

In collections Indian Museum and de Nicéville.

GENUS XXXVII.—CASYAPA.

Casyapa, Kirby, Syn. Cat. Diurn. Lep., p. 576 (1871).

Chætocneme,* Felder, Sitzb. Ak. Wiss., Math. Nat. Cl., vol. xl, p. 460 (1860).

Casyapa, Distant, Rhop. Mal., p. 385 (1886).

"In this genus the anterior wings are relatively somewhat shorter and broader than in the preceding genera (*viz., Baoris, Telicota, Satarupa*), the costal margin is slightly falcate at apex, the outer margin nearly straight; the upper disco-cellular nervule is shorter than the lower, which is obliquely directed inwardly, and the base of the second median nervule is a little more than twice as far apart from that of the lower as from that of the upper median nervule. The posterior wings are subovate, the first and second median nervules having an apparently common origin at about end of cell. The body is robust and hairy, the palpi broad, thickly clothed with somewhat short hairs and directed upwards and forwards; the antennæ are of moderate length, with a well-thickened curved club, which is not so strongly hooked as in *Satarupa*; the posterior tibiæ are very prominently spined and clothed with very long hairs." (*Distant, l. c.*)

---

* "This name was preoccupied by *Chætocneme* in Coleoptera and therefore the genus was rightly renamed by Mr. Kirby." (*Distant, l. c.*)

157.—CASYAPA PHANÆUS, *HEWITSON.*

Eudamus phanœus, Hewitson, Desc. Hesp., p. 14, n. 24 (1867).
Casyapa phanœus, Distant, Rhop. Mal., p. 386, pl. xxxv, fig. 18 (1886).
Erionota ? lalita, Doherty, J. A. S. B., 1886, p. 263.

" ♂. Above light but very bright ferruginous, slightly paler outwardly, marked with translucent ochreous spots set in black rings. On the forewing, one large triangular spot near the end of the cell, another larger and quadrate below it between the lower median branches, a third much smaller and rounded, slightly beyond them between the upper medians. Below these two minute ones with the translucent pupil obsolescent, set obliquely in the interno-median space : also five subapical ones, small, well separated, the third and fourth furthest from the base, the third largest, the fourth and fifth minute. Hindwing with a transverse black spot at the end of the cell, and a semicircle of eight smaller ones on the disc round it, of which only one or two are pupilled with hyaline: the first which is placed basally between the costal and subcostal veins is obscure. Below duller ferruginous, the markings similar, the black rings of the spots of the forewing less distinct. Body densely clad with rust-red fur, eyes scarlet. Expanse sixty millimetres. ♀ unknown, probably very similar.

I caught two males on Sirtai Mountain (2,000 ft.), in the Lushai country, north-east of Chittagong and South of Cachar.

The type of *Erionota* is *thrax,* but Mabille includes *irava* in the genus. On account of its possible affinity with that species (which I have never seen), I put this rare butterfly under the head of *Erionota.* It seems however to bear more resemblance to Felder's species of *Chœtocneme* and *Netrocoryne* from the Australian and Austro-Malayan region." (*Doherty, l. c.*)

This species was first described from Sarawak, Borneo, by Hewitson. Distant records it from Malay Peninsula, Singapore (*Wallace, coll. Godman and Salvin*).

In collection de Nicéville, and there are several specimens in the Rangoon Museum, from Rangoon, and also from Myittha in Tavoy.

158.—CASYAPA LIDDERDALI, *ELWES.*

Chœtocneme ? lidderdali, Elwes, Trans. Ent. Soc. Lond., 1888, p. 459.

"This remarkable species is only known to me from a single specimen in the British Museum, which came out of Lidderdale's collection, and though it may possibly have come from Buxa, is more probably a Sikkim insect.

Colour olive-brown, darker towards the apex, with yellowish-olive hair on thorax and hindwing. A series of irregular transparent spots near the apex of forewing, and five larger ones in a band across it, the largest of which closes the end of the cell. On the hindwing above is a series of eight oblong black spots margined with light olive. Fringe of hindwing light olive. Beneath the markings are similar, with the addition of a black oblong spot across the end of the cell of the hindwing. Abdomen dark, with olive bands. Antennæ brown.

Expanse : 2 inches.

This species is placed without a name in the genus *Chætocneme* in the British Museum collection.

A species which seems allied to this has recently been described from the Tipperah Hills by Doherty in J. A. S. B., 1886, p. 263, as *Erionota lalita.*" (*Elwes, l. c.*)

GENUS XXXVIII.—GANGARA.

Gangara, Moore, Lep. Cey., vol. i, p. 164 (1881).
Gangara, Distant, Rhop. Mal., p. 394 (1886).

" Wings, large ; forewing elongated, triangular ; costa arched at the base ; apex bluntly pointed ; exterior margin short, oblique ; cell broad, clavate, extending two-thirds the wing ; subcostal much arched along the cell, its branches at equal distances apart, first branch at two-fifths before end of the cell ; disco-cellulars inwardly oblique, upper bent near subcostal, upper radial from its angle, lower from their middle ; the middle median at one-sixth and lower at four-sixths before end of the cell ; submedian curved in the middle : hindwing short, broad, somewhat quadrate ; apex convex, exterior margin somewhat produced and convexly angular in the middle ; abdominal margin short ; cell short, extending one-third the wing and of equal width throughout ; costal vein very slightly arched, second subcostal at one-third before end of the cell ; disco-cellulars obliquely concave, slender ; no radial visible ; middle median from near end and lower at one-half before end of the cell ; both the middle and lower medians and the submedian vein in the male swollen towards the base, and clothed on the upperside of the wing with long hair; submedian and internal straight. Body large,.robust ; palpi thick, compactly flattened outwardly in front, terminal joint very short, broad, conical ; antennæ slender, tip finely pointed.

Type : *G. thyrsis.*" (*Moore, l. c.*)

159.—GANGARA THYRSIS, *FABRICIUS*.

Papilio thyrsis, Fabricius, Syst. Ent., p. 532 (1775).
Telegonus thyrsis, Butler, Cat. Lep. Fab., p. 262.
Hesperia pandia, Moore, P. Z. S., 1865, p. 790.
Gangara thyrsis, Moore, Lep. Cey., vol. 1, p. 163, pl. 66, figs. 3, 3 *a*, (1881).
Gangara thyrsis, Distant, Rhop. Mal., p. 394, pl. 35, fig. 13 (1886).

" Male and female dark chocolate-brown. Forewing with bright yellow semi-transparent quadrate spots disposed triangularly, the first large

---

* *Vide* preceding species.

and occupying half the cell, the second also large, obliquely beneath and partly beyond, the third small and obliquely above the second; above the last are three smaller spots obliquely before the apex, the two upper being geminated; in some specimens beneath the subapical spots is a small dot, and on the posterior margin another, both similar to the rest; cilia at posterior angle brownish-white; hindwing with the cilia at the anterior angle brownish-white. Underside, forewing irrorated with grey scales near the apex, posterior margin pale brownish-white, spots yellow as above; hindwing irrorated with grey scales in a series of bands across the wing.

Expanse: $2\frac{5}{8}$ to $3\frac{1}{4}$ inches.

Habitat: Bengal." (*Moore, l. c.* in P. Z. S.)

"Larva greyish-white with a few ochreous dorsal spots and marks. From the body, according to Dr. Thwaites, a loose shaggy filamentous clothing consisting of pure wax is excreted, but which is easily rubbed off when handled, leaving the larva quite naked. Feeds on *Palmaceæ*. Pupa pale olivaceous-yellow; the tongue spirally protruded." (*Moore, Lep. Cey., l. c.*)

Recorded from Ceylon (*Wade*; *Mackwood*); Bombay (*Swinhoe*); Nicobars, Andamans, and Cachar (*Wood-Mason* and *de Nicéville*); Calcutta (*de Nicéville*); Orissa (*Taylor*); Nilgiris (*Hampson*).

"The male of this species presents, on the upperside of each anterior wing, three lines of modified scales, namely, one along the posterior side of the median vein between the origins of its first and second branches, another on each side of the first median veinlet from the origin of this up to the second discal spot, and a third, also double, along an equal portion of the submedian vein, and a thick clothing of setæ paler than the ground-colour at the base of the interno-median area, and a similar clothing of paler setæ on the middle three-fourths of the sutural area; and, on the underside, a conspicuous and equally long furry patch of pale fulvous coarse setæ divided by the submedian vein." (*Wood-Mason* and *de Nicéville, J. A. S. B.,* 1881, p. 261).

This species occurs commonly at Rangoon and Beeling, Upper Tenasserim.

In collections Indian Museum and de Nicéville.

GENUS XXXIX.—HIDARI.

Hidari, Distant, Rhop. Mal., p. 395 (1886).

"This genus differs from *Erionota* and *Gangara* in having the upper disco-cellular nervule of the anterior wings longer than the lower; the base of the second median nervule of the anterior wings is also not more than twice as far apart from lower as from upper median nervule." (*Distant, l. c.*)

160.—HIDARI IRAVA, *MOORE*.

Hesperia irava, Moore (Horsfield and Moore), Cat. Lep. Mus. E. I. C., vol. i, p. 254, n. 583 (1857).

Hesperia irava, Plötz, Stett. Ent. Zeit., vol. xliii, p. 328, n. 80 (1882).

Celænorrhinus thrax, Hübner (*nec* Linnæus), Zutr. Ex. Schmett., figs. 875, 876 (1832).

Hesperia hypœpa, Hewitson, Desc. Hesp., p. 25, n. 7 (1868).

Erionota hypœpa, Mabille, Ann. Soc. Ent. Belg., vol. xxi, p. 35, n. 115 (1878).

Casyapa irava, Butler, Trans. Linn. Soc., second series, Zool., vol. i, p. 553, n. 2 (1877).

Erionota irava, Mabille, Ann. Soc. Ent. Belg., vol. xxi, p. 35, n. 116 (1878).

Hidari irava, Distant, Rhop. Mal., p. 395, pl. xxxiv, fig. 15, ♀ (1886).

"Male and female. Wings above chocolate-brown; anterior wings with the basal costal area rufous, and the disc much darker, containing four pale ochraceous spots, situate one in cell, and three beneath cell divided by the median nervules, and two (sometimes only one) small greyish subapical spots; posterior wings with the fringe greyish-ochraceous. Wings beneath very much paler than above; anterior wings with the disc blackish, spotted as above, and with small fuscous spots divided by the lower discoidal nervule; posterior wings with usually four small fuscous discal spots in curved series. Body and legs more or less concolorous with wings.

Expanse: ♂ and ♀, 52 to 64 millims.

Habitat: Malay Peninsula, Penang (*coll. Distant*); Perak (*Künstler, coll. Distant*); Malacca (*coll. Staudinger*; *Pinwill, Brit. Mus.*; *Biggs, coll. Distant*); Java (*Horsfield*)." (*Distant, l. c.*).

In collections Indian Museum and de Nicéville, and there is a single specimen in the Rangoon Museum from Myittha, Tavoy.

161.—HIDARI BHAWANI, *DE NICÉVILLE*.

Hidari bhawani, de Nicéville, J. A. S. B., vol. lvii, pt. 2, p. 291, n. 23, pl. xiii, f. 6, ♂ (1888).

"Male. Upperside, both wings ochreous-brown. Forewing with four lustrous semi-transparent pale yellow spots, one just beyond the middle of the cell much constricted in the middle, an oval one in the upper discoidal interspace, a squarish one near the middle of the second median interspace, and the last near the middle of the first median interspace, lunular; a small opaque spot in the submedian interspace touching the middle of the submedian nervure. Hindwing unmarked, but densely woolly towards the base. Underside, forewing brown, the costa and the apex broadly pale ochreous more or less striated with fine brown lines; the four semi-transparent spots as above, but with two minute ones above the subapical spot divided by the fourth subcostal nervule; the spot in the submedian interspace larger and diffused. Hindwing pale ochreous, but with a dark brown streak parallel and near to the costa from the base to the outer margin, and the abdominal margin widely brown, the ochreous portion of the wing coarsely striated with brown. Head and thorax above clothed with long pale ochreous hairs, but with a line of dark brown hairs running down the middle; abdomen dark brown above; palpi, thorax, and abdomen pale ochreous beneath; antennæ with the shaft pale ochreous above, dark brown beneath, club pale ochreous anteriorly, fuscous posteriorly.

Expanse: ♂ 2·2 inches.

Habitat: Arracan Coast, Burma.

Described from a single specimen in Major C. T. Bingham's collection, taken by him in February, 1886. It cannot be mistaken for the other three species of the genus, *H. irava*, Moore, *H. sybirita*, Hewitson, or *H. staudingeri*, Distant, all of which occur in the Malay Peninsula." (*de Nicéville, l. c.*)

GENUS XL.—PLASTINGIA.

Plastingia, Butler, Ent. Month. Mag., vol. vii, p. 95 (1870).

Plastingia, Distant, Rhop. Mal., p. 396 (1886).

"This genus has the upper and lower disco-cellular nervules of the anterior wings almost subequal in length, thus agreeing with *Erionota* and *Gangara*, but easily distinguished in a structural sense from those genera by the position of the second median nervule of the anterior wings, which has its base about three times as far apart from that of the lower as from that of the upper median nervule. Another distinguishing character in *Plastingia* is the position of the first subcostal nervule of the anterior wings, which is emitted more nearly opposite the base of second than that of lower median nervule." (*Distant, l. c.*)

162.—PLASTINGIA CALLINEURA, *FELDER*.

Hesperia callineura, Felder, Reise Nov., Lep., vol. iii, p. 513, n. 895, pl. 71, figs. 9, 10 (1866).

Plastingia callineura, Druce, Proc. Zool. Soc., 1873, p. 359, n. 2.

Plastingia callineura, Butler, Trans. Linn. Soc., Second Series, Zool., vol. i, p. 355, n. 1 (1877).

Plastingia callineura, Mabille, Ann. Soc. Ent. Belg., vol. xxi, p. 36, n. 132 (1878).

Plastingia callineura, Plötz, Stett. Ent. Zeit., vol. xlv, p. 148, n. 12 (1884).

Hesperia latoia, Hewitson, Descr. Hesp., p. 34, n. 27 (1868).

Hesperia latoia, Hewitson, Ex. Butt., *Hesperiidæ*, pl. 6, figs. 62, 63 (1873).

Plastingia latoia, Plötz, Stett. Ent. Zeit., vol. xlv, p. 149, n. 13 (1884).

Plastingia callineura, Distant, Rhop. Mal., p. 396, pl. xxxv, fig. 26 (1886).

"Wings above dark chocolate-brown; anterior wings with basal, costal, subcostal, and inner-marginal ochraceous streaks, and with a discal series of seven pale semi-hyaline spots, of which the largest is bifid and situate above the lower median nervule, three—the uppermost smallest—in oblique series separated by the lower discoidal and upper median nervules, two small and subapical, and one in and near end of cell; posterior wings with a central transverse fascia connected with base, a narrow streak along inner edge of abdominal margin, and the fringe ochraceous. Wings beneath paler than above; anterior wings spotted as above, and with a series of dark submarginal streaks placed between the nervules; posterior wings with the neuration ochraceous, the discal fascia as above, and with a submarginal series of minute ochraceous spots. Body above brownish, with segmental ochraceous fasciæ; body beneath pale ochraceous, abdomen with a central series of dark brownish spots; legs brownish.

Expanse: 38 millims.

Habitat: Malay Peninsula; Malacca (*Pinwill, Brit Mus.*); Singapore (*coll. Hewitson; Wallace, coll. Godman* and *Salvin*); Java, Buitenzorg (*Felder*).

A single specimen of this species has been obtained in Mergui by Dr. Anderson.

In collection Indian Museum.

163.—PLASTINGIA NOËMI, *DE NICÉVILLE.*

Plastingia noëmi, de Nicéville, J. A. S. B., vol. liv, pt. 2, p. 120, pl. ii, fig. 15 ♂ (1885).

" ♂. Upperside black. Forewing with a fusiform chrome-yellow streak on the costa from the base to nearly half the length of the wing, a similarly-coloured streak placed below the median nervure and divided by the submedian into two unequal parts, the lower portion the smaller, extending to rather more than half the inner margin of the wing from the base; and with two or three subapical conjoined increasing spots, two lengthened spots at the end of the cell placed one above the other, the upper one the smaller, a triangular spot towards the base of the second median interspace, a much larger one towards the base of the first, all semi-transparent yellowish-white. Hindwing with a chrome-yellow patch placed in the middle of the disc just beyond the cell, and divided by the black nervules. Underside, forewing black, the costa narrowly, the apex very widely, and a patch placed in the middle of the submedian interspace chrome-yellow. The semi-transparent spots as above. Five rounded small black spots placed in an outwardly-angled subapical series. Hindwing chrome-yellow; the margin increasingly to the anal angle, then decreasingly up the abdominal margin black. A subbasal spot, another at the end of the cell, a series of eight spots placed round the cell, all black. Antennæ black, the club yellow. Thorax and base of abdomen above clothed with long greenish-ochreous hairs, the rest of the abdomen black ringed with yellow, the thorax and legs beneath chrome-yellow.

Expanse: 1·6 inches.

Habitat: Sikkim (*Otto Möller* and *Dr. T. C. Jerdon*).

Belongs to the same group as the *Hesperia callineura* of Felder (= *Hesperia latoia*, Hewitson) but is quite distinct." (*de Nicéville, l. c.*)

In collection de Nicéville.

164.—PLASTINGIA NAGA, *DE NICÉVILLE.*

Hesperia? naga, de Nicéville, J. A. S. B., vol. lii, pt. 2, p. 89, n. 37, pl. x, fig. 2 ♀ (1883).

"Female. Upperside, both wings brown; the cilia cinereous, dark brown at the end of the nervules. Forewing with a spot at the end of the cell; two smaller spots beyond, the lower one twice the size of the upper; an elongated spot near the middle of the second median interspace, and another, the largest of all, near the base of the first median interspace; all these spots semi-transparent ochreous-white. A subcostal narrow yellow streak extending from the base to beyond one-third of the length of the wing, a similar one touching and placed above the submedian nervure, extending from the base to beyond half the length of the inner margin of the wing. Hindwing with an elongated streak of ochreous hairs in the cell, and a series of short ochreous streaks between the nervules placed outside it; a similar streak extending from the base to near the margin and touching the inner side of the submedian nervure. Cilia alternately cinereous and dark brown. Underside, both wings lighter brown; the cilia white, brown at the end of the nervules. Forewing with the spots as above but whiter, and edged with pure white; a subcostal streak extending from the base to nearly half the length of the wing, broadest at its end; beyond which are some streaks between the subcostal branches, two similar streaks in the discoidal interspaces, and a marginal series ending at the first median nervule; the two middle spots small; a wide streak extending to beyond the middle of the wing from the base placed in the submedian interspace:—all pure silvery white. Hindwing marked with about eighteen silvery white spots and streaks disposed equally over the whole surface of the wing. Body brown, the thorax thickly clothed with long ochreous hairs, the abdominal segments ringed with ochreous, paler below.

A single specimen has been obtained by Mr. S. E. Peal.

Expanse: ♀, 1·6 inches.

Habitat: Sibsagar, Upper Assam." (*de Nicéville, l. c.*)

The male is at present unknown.

In collection Indian Museum.

165.—PLASTINGIA MARGHERITA, *DOHERTY.*

Plastingia margherita, Doherty, J. A. S. B., 1889, p. 131, pl. x, fig. 5, ♂.

"Male. Above black, with light golden-ochreous translucent markings, and richer orange-ochreous opaque ones. Of the former there are on the forewing, two unusually large, elongate-quadrate, subapical ones, separated by a vein, the lower longest; one large oblique cellular one of hour-glass shape; and three discal ones in echelon, of which one is very large, occupying the entire breadth of the lower median space, irregularly pentagonal, twice as long as broad, separated from the cell-spot only by the black median vein; the other two smaller, elongate, broadest outwardly. Also with the following opaque markings:—one above the cell and one in the interno-median space, extending obliquely from the internal vein, not far from the base, to the lower median vein, which separates it from the basal part of the larger discal spot. Hindwing with a large irregular ochreous patch in the disc just beyond the cell, consisting of two translucent areas joined by the opaque orange-ochreous base of the upper median space, the outer one larger, obliquely quadrate, between the lower subcostal and upper median branches, the other occupying the basal part of the lower median space. Below blackish, the veins, except near the abdominal margin of both wings, widely bordered with reddish-ochreous. Forewing with the rufous costal area extending over the upper part of the cell; that in the interno-median space much larger and paler than above. Hindwing with a number of lustrous lilac markings in the black spaces between the reddish nerve-rays, namely, two in the cell, the basal one elongate, one at the base of the costa, elongate, two in the upper subcostal space, the outer one elongate, one in the lower subcostal space, quadrate, and three in the median and submedian spaces, in a line receding from the border. Cilia cinereous.

One male, Margherita, and a similar one, Sadiya.

The species is a local form (differing in the large subapical spots, the absence of the outer-fourth discal spot, the undivided cell-spot separated from the interno-median one, and in the ochreous patch of the hindwing consisting of two hyaline and one opaque space and confined to the disc) of another found in the three Indo-Malayan islands, the Malayan Peninsula, and Mergui, but everywhere rare. The Javanese form (*callineura*) seems, judging by my specimens, to differ but slightly. The single, very worn Mergui specimen, taken by Dr. Anderson, has been identified by Mr. Moore as *Plastingia latoia*, Hewitson. But that species (and *P. callineura*, Felder, which is regarded as conspecific with it) has been described and figured by Hewitson, Felder and Distant with ochreous submarginal spots on the hindwing below, no blue ones being mentioned. In any case the above mentioned characters separate my species as a distinct local form.

The egg of several species of *Plastingia* examined by me, generally resembles that of *Suastus*. But like those of *Hesperia satwa*, de Nicéville, and the species of *Cupitha*, though in a lesser degree, it possesses a large crown-like mass of white cells apically, surrounding the micropyles, as delicate in structure as the finest lace. They are the most beautiful butterfly-eggs known to me." (*Doherty, l. c.*)

GENUS XLI.—HYAROTIS.

Hyarotis, Moore, Lep. Cey., vol. i, p. 174 (1881).
Hyarotis, Distant, Rhop. Mal., p. 397 (1886).

" Forewing triangular; costa long, apex pointed, exterior margin very oblique, posterior margin short in male; subcostal at equal distances apart, first branch at one-third before end of the cell; the cell two-thirds the length of the wing; disco-cellulars inwardly oblique; upper radial from angle near subcostal, lower from their middle; the middle median at one-seventh, lower at four-sevenths before end of the cell; submedian slightly arched in the middle: hindwing oval; exterior margin convex; cell extending half the wing, broad; disco-cellulars very slender; radial very indistinct; middle median from close to end, and lower at one-third before end of the cell; submedian straight; internal slightly curved. Body moderate; palpi flattened laterally in front, terminal joint short, conical, pointed; antennæ slender, with a stoutish club and slender tip.

Type: *H. adrastus*." (*Moore, l. c.*)

166.—HYAROTIS ADRASTUS, *CRAMER*.

Hesperia adrastus, Cramer, Pap. Exot., vol. iv, pl. 319, figs. F, G (1780).
Plesioneura praba, Moore, P. Z. S., 1865, p. 790.
Hesperia phænicis, Hewitson, Exot. Butt., Hesp., pl. 4, figs. 36, 37 (1869).
Hyarotis adrastus, Moore, Lep. Cey., vol. i, p. 174, pl. 67, figs. 5, 5 *a* (1881).
Hyarotis adrastus, Distant, Rhop. Mal., p. 397, pl. xxxiv, fig. 4 (1886).

" Male and female, dark chocolate-brown. Upperside, forewing with three small conjugated subapical semi-transparent white spots, three similar and larger discal spots, and a fourth above them within the cell. Underside darker brown basally, paler exteriorly; forewing with spots as above, bordered externally by a suffused dark brown streak; hindwing with a double series of white dark brown-outer-bordered lunules crossing the middle of the wing, beyond which is a submarginal series of suffused dark brown spots. Palpi, thorax, and abdomen beneath pale greyish-brown. Legs brown. Cilia yellowish-white, spotted with pale brown.

Expanse: ♂ $1\frac{3}{8}$; ♀ $1\frac{5}{8}$ inches.

Habitat: Bengal." (*Moore, l. c. in P. Z. S.*) Recorded from Ceylon (*Hutchison, Wade, Mackwood*); Andamans, Cachar (*Wood-Mason* and *de Nicéville*); Sikkim (*de Nicéville; Elwes*); Calcutta (*de Nicéville*); Kumaon (*Doherty*); Kangra, N.-W. Himalayas (*Moore*); Orissa (*Taylor*); Nilgiris (*Hampson*).

I have obtained this species at Pegu and Rangoon.
In collections Indian Museum and de Nicéville.

GENUS XLII.—COLADENIA.

Coladenia, Moore, Lep. Cey., vol. i, p. 180 (1881).
Coladenia, Distant, Rhop. Mal., p. 397 (1886).

"Wings short: forewing triangular, apex pointed, exterior margin short, slightly oblique and waved; cell two-thirds the length; first subcostal at less than one-third before end of the cell, first, second and third at equal distances, fourth and fifth from end of the cell; disco-cellulars slightly inwardly oblique, upper radial from angle near subcostal, lower from their middle; the middle median at one-sixth, lower at four-sixths before end of the cell; submedian recurved: hindwing broad, somewhat quadrate; exterior margin slightly waved and angular in the middle; cell broad, extending more than half the wing; second subcostal at one-fourth before end of the cell; disco-cellulars nearly straight, upper shortest, radial from their middle; the middle median from immediately before end of the cell, lower at nearly one-half before the end. Body rather stout; palpi of lax hairy scales: terminal joint short, thick, cylindro-conical; antennæ long with gradually thickened club and slender tip; fore tibiæ tufted beneath; femora pilose beneath, hind tibiæ with a long tuft of hair above.

Type: *C. indrani*." (*Moore, l. c.*)

167.—COLADENIA INDRANI, *MOORE*.

Plesioneura indrani, Moore, P. Z. S., 1865, p. 789.

"Upperside bright golden-yellow: forewing with a discal series of four semi-transparent white black-bordered spots, the first small and above the extremity of the cell, the second large quadrate and within the cell, the third elongate and beneath the latter, the fourth exterior to their juncture: beneath these is a pale golden-yellow black-bordered spot: before the apex is a series of four similar white spots with black border, the three upper of which are conjugated; a well-defined black spot beneath the cell near the base; exterior margin and cilia blackish, the latter white at the posterior angle: hindwing with a semi-circular submarginal series of black spots and two similar inner discal spots; exterior margin black; cilia alternate black and white. Underside blackish-brown suffused with golden-yellow, brightest on the hindwing: markings as above but more clearly defined. Tip of palpi black: thorax, body, palpi (except tip), and legs yellow. Antennæ yellow, tip black.

Expanse: 1¾ inches.
Habitat: Bengal." (*Moore, l. c.*)

Recorded from Mhow (*Swinhoe*); Orissa (*Taylor*); Sikkim (*de Nicéville*; *Elwes*).

In collections Indian Museum and de Nicéville.

168.—COLADENIA TISSA, *MOORE*.

Coladenia tissa, Moore, Lep. Cey., vol. i, p. 180, pl. 67, fig. 6 (1881).

"Male. Upperside brownish-ferruginous, with indistinctly paler marginal lunules: forewing with three or four very small semi-transparent yellowish-white subapical spots, a transverse medial series of larger spots composed of a large irregular-shaped spot at end of the cell, a smaller spot on the costa above it, a large spot beneath it below end of the cell, and a very small spot between them at the junction of upper and middle median veins; between these is an irregular triangularly-lobed black-bordered ferruginous spot; a small black spot below the cell near the base; hindwing with an indistinct blackish discal spot, and a discal series of spots curving from base of costa to above anal angle. Cilia brownish-ferruginous, with a slight paler streak above posterior angle of forewing. Underside dusky ferruginous-brown; markings more prominent. Palpi and legs pale ferruginous. Female paler, with marginal lunules more diffused; markings as in male.

Expanse: ♂ 1⅜; ♀ 1⅞ inches. *Terai, Sikkim.*

Allied to *C. indrani*. Of a darker colour; has deeper yellow and differently-shaped semi-transparent spots on forewing, and the cilia is not alternated with white." (*Moore, l. c.*)

Occurs Ceylon (*Wade, Mackwood*); Calcutta (*de Nicéville*); Nilgiris. (*Hampson*).

Mr. de Nicéville notes that in Calcutta there are two broods, the one that appears in the rains differing from the cold-weather generation in having the ground-colour of both wings umber-brown, instead of ochreous, and all the black spots and markings more prominent.

In collections Indian Museum and de Nicéville.

169.—COLADENIA FATIH, *KOLLAR*.

Hesperia fatih, Kollar, Hügel's Kaschmir, vol. iv, pt. 1, p. 454, n. 3, pl. xviii, figs. 5, 6 (1848).

"Expanse: 1″ 5¼.‴ (1.75 inches in figure).

Wings yellow-brown; in the middle of forewing four spots, two adjacent dots and three lesser spots at the apex, transparent white, on the hindwing two dusky obsolete bands.

Of the size of *Hesperia tessellum*; wings dark brown covered with yellowish-grey scales so that the ground-colour is only visible here and there; on the middle of the forewing are two big quadrangular white transparent spots and between them a third smaller one, on the costal margin but in the same line there is a fourth still smaller one, and besides beyond *the second window-like spot there are also two transparent dots; lastly there are three similar transparent spots before the apex arranged in a semi-circle. On the hindwing the ground-colour is visible as two indistinct badly-defined bands. The underside is duller, but otherwise agrees entirely with the upperside. Fringe alternately brown and white. Head and thorax clothed with reddish-yellow hairs, antennæ brown.

Hügel obtained a single example in the Himalayas." (*Kollar, l. c.*)

In collection Indian Museum.

---

* *i. e.*—between it and the inner margin.

170.—COLADENIA DAN, *FABRICIUS*.

Papilio dan, Fabricius, Mant. Ins., vol. ii, p. 88, n. 798 (1787).
Coladenia dan, Distant, Rhop. Mal., p. 398, pl. xxxv, fig. 27 (1886).
Plesioneura dan, var. *andamanica*, Wood-Mason and de Nicéville, J. A. S. B., vol. l, pt. 2, p. 257, n. 118 (1881); J. A. S. B., vol. lv, pt. 2, p. 391, n. 246 (1887).

" Wings above rufous-brown; anterior wings with four discal greyish-white spots, one—largest and sublunate—in cell, a small spot above it, and two beneath cell divided by the second median nervule, and three—sometimes two—small subapical greyish-white spots in suberect series, outer half of wing with obscure dark fasciæ; posterior wings with discal and outer marginal dark fasciæ. Wings beneath as above, but slightly paler. Body and legs more or less concolorous with wings.
Expanse : 30 millim." (*Distant, l. c.*)

Recorded from Kumaon (*Doherty*); Sikkim (*de Nicéville, Elwes*); Cachar (*Wood-Mason* and *de Nicéville*); Kangra, N.-W. Himalayas (*Moore*); Nilgiris (*Hampson*).

I have obtained this species commonly in Rangoon; Beeling, Upper Tenasserim; the Nilgiris; and the Kadur District, Mysore. Mr. de Nicéville considers that *C. fatih* is distinct from this species though the general opinion seems to be they are identical. I am unable to say by what characteristic they are to be separated, though *C. fatih* seems to be constantly larger.

The following is the description of the local race *andamanica*:—

" Specimens from Port Blair differ from continental ones in having the discal series of spots all run together so as to form an unbroken band, and the three subapical spots conjugated and in the same straight line, with their conjoined inner margin nearly straight and their outer festooned; and two examples exhibit in addition two smaller dots placed nearly parallel to the outer margin just below and external to the three subapical ones." (*Wood-Mason* and *de Nicéville l.c. in J. A. S. B., vol. l.*)

Habitat: Andamans, Cachar.

In recording this variety from Cachar, Messrs. Wood-Mason and de Nicéville note that specimens from that locality differ from Andaman specimens in the discal golden band of the upperside of the forewing being slightly broader, and the yellow spots on the underside of the hindwing obsolete; and that it agrees with *P. dhanada*, Moore, in the golden band not nearly reaching the anal angle of the forewing on the upperside (thereby differing from *P. aurivittata*, Moore), departing from it in having the cilia cinereous throughout (*J. A. S. B., vol. lv*).

In collection Indian Museum.

171.—COLADENIA HAMILTONII, *DE NICÉVILLE.*

Coladenia hamiltonii, de Nicéville, J. A. S. B., vol. lvii, pt. 2, p. 291, n. 24, pl. xiii, fig. 8, ♂ (1888).

" Male. Upperside, forewing olive-greenish fuscous, with two very irregular broad discal black fasciæ joined in the middle; three most minute transparent subapical dots, the uppermost the largest, placed at the outer edge of the anterior portion of the outer black fascia, a very minute similar spot in the second median interspace, a very attenuated spot across the middle of the first median interspace, both placed on the outer black fascia; the inner margin somewhat broadly irrorated with greyish scales; a submarginal indistinct broad blackish fascia. Hindwing, ground-colour much as in the forewing, but the outer third of the wing irrorated with grey scales; a recurved black macular decreasing band from the costa near the apex of the wing to the second median nervule; the disco-cellular nervules defined by a pale line. Underside, both wings vinous-fuscous. Forewing with the transparent spots as above. Hindwing with the disc irrorated with whitish; the macular black band much as above; an anteciliary whitish line. Cilia fuscous.

Expanse : ♂ 1·6 inches.
Habitat : Sylhet.

The Rev. Walter A. Hamilton, after whom I have much pleasure in naming it, obtained a single specimen in Sylhet in the spring. It is quite unlike any species known to me, and I place it in the genus *Coladenia* only because it agrees in outline with *C. tissa*, Moore." (*de Nicéville, l. c.*)

In collection de Nicéville.

GENUS XLIII.—TAPENA.

Tapena, Moore, Lep. Cey., vol. i, p. 181 (1881).

" Wings small; forewing short, apex acute; exterior margin slightly oblique, convexly angular in the middle; first subcostal emitted at nearly one-half before end of the cell; cell extending two-thirds the wing disco-cellulars obliquely convex, upper bent close to subcostal, upper radial from its angle, lower from their middle; the middle median branch at one-fifth, lower at four-fifths before end of the cell; submedian slightly curved in the middle; hindwing quadrate, short, broad; apex slightly and middle of exterior margin acutely angular; abdominal margin somewhat long; cell broad, extending nearly two-thirds the wing; second subcostal at one-third before end of the cell; disco-cellulars slender, outwardly oblique, radial very slender, middle median from close to end of the cell; lower at one-fourth before the end; submedian straight, internal curved. Body robust; palpi laxly compressed, terminal joint rather long, thick, cylindrical; antennæ slender, somewhat short, hooked tip slender.

Type: *T. thwaitesi.*" (*Moore, l. c.*)

172.—TAPENA THWAITESI, *MOORE*.

Tapena thwaitesi, Moore, Lep. Cey., vól. i, p. 181, pl. 67, figs. 2, 2a (1881).

"Male. Upperside dark purple-brown, with blackish outer margins and indistinct discal transverse macular fascia: forewing with two small semi-diaphanous white spots on the costa near the apex. Underside dark purple-brown. Female. Upper-and underside greyish purple-brown, transverse macular fascia and outer border dark purple-brown: forewing with three small semi-diaphanous white subapical spots, the lowest transversely narrow, two spots also at end of the cell, the upper one very slender, and two spots on the disc, each series being bordered by the dark fascia: hindwing with a semi-diaphanous spot at end of the cell.

Expanse: $1\frac{3}{10}$ to $1\frac{8}{10}$ inches." (*Moore, l. c.*)
Occurs Ceylon (*Wade*); Orissa (*Taylor*); Nilgiris (*Hampson*).
In collections Indian Museum and de Nicéville.

173.—TAPENA AGNI, *DE NICÉVILLE*.

Plesioneura agni, de Nicéville, J. A. S. B., vol. lii, pt. 2, p. 87, n. 32, pl. x, fig. 4, ♀ (1883).

Cælænorrhinus agni, de Nicéville, Journ. Bomb. Nat. Hist. Soc., vol. iv, p. 186, n. 31 (1889).

"Male. Upperside dark brown, but so thickly covered with large fulvous overlying scales as to leave the ground-colour visible only on the outer margin, a streak within the apical spots, and narrowly all round the transparent white spots. Forewing with a large quadrate spot filling the end of cell, a small spot above it, a rather larger one at the base of the second median interspace, a large one nearly equal in size to the spot in the cell at the base of the first median interspace, and two small rounded spots in the submedian interspace placed obliquely, the upper one below the outer lower angle of the spot above; three or four small subapical spots, the upper one rather larger than the rest, the second out of line, being placed nearer the base of the wing:—all these spots lustrous semi-transparent white. Cilia dark brown, with a pale spot at the apex, and another larger one on the submedian interspace. Hindwing with a black spot at the end of the cell (sometimes obsolete), and a curved series of eight similar spots, the two upper ones round, the others oblong and placed in pairs, the two lowest spots—as in the specimen figured—sometimes obsolete. Cilia dark brown, paler towards the apex. Underside pale brown. Forewing with the spots as above, but with a pale fulvous submarginal curved fascia. Hindwing as above, but paler fulvous, the spots more prominent. Female a little paler than the male, the spots somewhat larger. Body fulvous, antennæ black above, paler below.

Nearest to *P. chamunda*, Moore, but conspicuously differing from that species in having the hindwing marked with black spots above and below, and the cilia not alternately brown and white as in that species.

Expanse: ♂ 1·6; ♀ 1·8 inches." (*de Nicéville, l. c.*)

Habitat: Mr. de Nicéville records two pairs from low elevations in Sikkim, and informs me that it also occurs in the Khasi Hills, and I obtained a single specimen at Tilin in the Yaw District, Upper Burmah.

In collections Indian Museum and de Nicéville.

174.—TAPENA LAXMI, *DE NICÉVILLE.*

Plesioneura laxmi, de Nicéville, Journ. A. S. B., vol. lvii, pt. 2, p. 290, n. 21, pl. xiii, fig. 5, ♀ (1888).

Oelænorrhinus laxmi, de Nicéville, Journ. Bomb. Nat. Hist. Soc., vol. iv, p. 186, n. 32 (1889).

" Habitat: Upper Tenasserim.

Expanse: ♀ (nec. ♂), 1·8 inches.

Description [Female] : Upperside, *both wings,* olive-greenish. *Forewing* with a very large quadrate spot filling the outer end of the discoidal cell and extending somewhat narrowly to the costa; a small quadrate spot near the base of the second median interspace; another quadrate spot below it fully four times as large, in the middle of the first median interspace; two dots placed obliquely in the submedian interspace, the upper one placed below the lower outer angle of the large spot in the interspace above; three subapical well-separated dots in a curved series, the upper one twice as large as the other two taken together—all these spots lustrous transparent white; an indistinct dark macular submarginal band, and two small black dots placed one above the other obliquely near the base of the submedian interspace. *Hindwing* rather paler than the forewing; a subcostal black spot placed near the base of the wing, two parallel discal black macular bands. Underside, *both wings* ochreous-brown. *Forewing* marked as above. *Hindwing* with the bands broken up into spots and arranged thus :—a largish black spot in the discoidal cell, almost completely surrounded by a series of spots beginning with a moderate-sized one near the base of the subcostal interspace, a very large round one near its middle, then about eight small spots curving round to the base of the wing. *Cilia* brownish throughout. *Antennæ* ochreous-brown above, the hook black above, ochreous below. *Body* and *head* more or less concolorous with the wings above and below.

I have a single specimen taken by Major C. T. Bingham in March in the Thoungyeen Forests, Burma. It is nearest to *T. agni,* mihi, but the ground-colour of the upperside is entirely different, as are also many of the markings." (*de Nicéville, l. c.* in *J. A. S. B.*)

The type specimen was originally described as a male, but Mr. de Nicéville informs me it is a female, and that he has since received the true male, which he will shortly describe.

175.—TAPENA BUCHANANII, *DE NICÉVILLE.*

Oelænorrhinus buchananii, de Nicéville, Journ. Bomb. Nat. Hist, Soc., vol. iv, p. 187, n. 21, pl. B, fig. 2, ♀ (1889).

"Habitat: Upper Burma.

Expanse: ♀, 2·1 inches.

Description: Female. Very closely allied to *C. laxmi*, mihi, from which it differs in its considerably larger size. Upperside, *forewing* with the white discal band fully twice as wide, not divided into spots, extending uninterruptedly from the costa to the submedian nervure, its edges very irregular, its lower portion posterior to the first median nervule much narrower than the rest of the band; this species lacks the two small obliquely-placed black dots found near the base of the submedian interspace in *C. laxmi*. *Hindwing* instead of possessing two parallel discal macular black bands has a rounded black spot towards the end of the discoidal cell, and a discal series of six black spots, of which the anterior one is round and well-separated from the spot which follows it, the second spot is round, the next pair are the largest and elongated, and the last pair smaller but also elongated; *cilia* of hindwing anteriorly white, posteriorly dark brown. Underside, *both wings* with the same differences as above, but all the spots of the hindwing more prominent.

I believe this to be a species distinct from *C. laxmi*, though a single male of that species only is known, while *C. buchananii* is described from a single female. The difference in size is very considerable, and is more than is usually found in the opposite sexes of the *Hesperiidæ*, and the markings also shew marked differences. I have named it after its capturer, Mr. A. M. Buchanan, who obtained it in the Ruby Mine District, Upper Burma." (*de Nicéville, l. c.*)

With reference to the last paragraph of above description, Mr. de Nicéville informs me that his type of *C.* (= *T.*) *laxmi* was a female and not a male, so that *T. buchananii* is undoubtedly a good species.

GENUS XLIV.—UDASPES.

Udaspes, Moore, Lep. Cey., vol. i, p. 177 (1881).
Udaspes, Distant, Rhop. Mal., p. 399 (1886).

"Allied to *Plesioneura*: forewing less triangular, exterior margin more convex, posterior margin longer; cell narrow at end; disco-cellulars extending very obliquely inward; lower median branch at less than two-thirds before end of the cell: hindwing very broadly oval, extremely convex externally: abdominal margin short: cell much shorter, second subcostal nearer end of the cell; upper and middle medians both emitted from extreme end of the cell; lower median at less than one-third before the end. Body short; antennal club shorter, and thicker at tip.

Type: *Udaspes folus.*" (*Moore, l. c.*)

178.—UDASPES FOLUS, *CRAMER*.

Papilio folus, Cramer, Pap. Exot., vol. i, pl. 74, fig. 7 (1779).
Hesperia cicero, Fabricius, Ent. Syst., vol. iii, p. 338 (1793).
Udaspes folus, Moore, Lep. Cey., vol. i, p. 177, pl. 68, figs. 3, 3*a* (1881).
Udaspes folus, Distant, Rhop. Mal., p. 398, pl. xxxiv, fig. 3 (1886).

"Upperside dark sepia-brown: forewing with three connected semi-transparent yellowish-white spots before the apex, two below them nearer the outer margin, a smaller spot between the upper and middle median veins, two larger connected spots on the disc, and a large spot within end of the cell: hindwing with a large medial discal semi-transparent yellowish-white sinuous-bordered patch. Cilia alternated with white. Underside: forewing as above: hindwing with the white patch extending medially to base of the wing and traversed by a dark brown sinuous streak below end of the cell; outer margin of the wing suffused with greyish-white. Body palpi and legs beneath greyish-white.

Expanse: $1\frac{1}{4}$ to $1\frac{3}{4}$ inches." (*Moore, l. c.*)

Occurs Ceylon (*Moore*); Poona, Ahmednagar, Bombay (*Swinhoe*); Kangra, N.-W. Himalayas (*Moore*); Calcutta (*de Nicéville*); Kumaon (*Doherty*); Cachar (*Wood-Mason* and *de Nicéville*); Orissa (*Taylor*); Sikkim (*de Nicéville*; *Elwes*); Nilgiris (*Hampson*).

I have obtained this species commonly at Poungadaw, Upper Burma; Beeling, Upper Tenasserim; and at Rangoon.

In collections Indian Museum and de Nicéville.

GENUS XLV.—NOTOCRYPTA.

Notocrypta, de Nicéville, Journ. Bomb. Nat. Hist. Soc., vol. iv, p. 188 (1889).
Plesioneura, Felder, Wien. Ent. Monatschr., vol. vi, p. 29 (1862), preoc.
Plesioneura, part, auctorum.

"Differs from *Cœlænorrhinus*, Hübner, in the forewing being more triangular, the *middle disco-cellular* nervule being distinctly longer instead of shorter than the lower disco-cellular, concave instead of almost straight, the middle and lower disco-cellular nervules taken together less strongly inwardly oblique; the hindwing is also shorter and more produced posteriorly, the *costa* is more arched, the *discoidal cell* is distinctly shorter, thus causing all the veins which spring from it (the first and second subcostal, the discoidal, and the three median nervules) to be distinctly longer.

There is a marked difference in the length of the haustellum or tongue, which in *C. leucocera*, Kollar, measures 1·8 inches, in *N. alysos*, Moore, only ·9 of an inch, or exactly half. Type, the *Plesioneura curvifascia* of Felder.

This diagnosis has been drawn up from bleached wings of both sexes of *N. alysos*, Moore, from Sikkim. All the species of this genus settle with closed wings; through an unfortunate and stupid mistake I once stated that they rest with wide outspread wings. This marked characteristic in life, which at once distinguishes *Notocrypta* from *Celænorrhinus*, has led me to discriminate these two genera; there is also considerable difference in the outline of the wings, and I believe *Notocrypta* never has the hindwing spotted, except in *N. paralysos*, Wood-Mason and de Nicéville, this being always a feature in *Celænorrhinus*. The type species, *N. curvifascia*, was described from China, and has been identified by Messrs. Plötz, Doherty, and Leech as synonymous with *N. alysos*, Moore, but an actual comparison of specimens is desirable. The transformations of *N. alysos*, Moore, only are known." (*de Nicéville, l. c.*)

177.—NOTOCRYPTA ALYSOS, *MOORE*.

Plesioneura alysos, Moore, P. Z. S., 1865, p. 789.

Plesioneura alysos, Moore, Lep. Cey., vol. i, p. 178, pl. 67, figs. 3 *a*, *b* (1881).

Plesioneura alysos, Distant, Rhop. Mal., p. 399, pl. 34, fig. 7 (1886).

Notocrypta alysos, de Nicéville, Journ. Bomb. Nat. Hist. Soc., vol. iv, p. 189, n. 2 (1889).

" Upperside dark fuliginous-brown; forewing with a broad oblique discal irregular-margined semi-transparent white band, and with one or two, and in some specimens three, very small similar conjugated spots obliquely before the apex, also one or two reversely oblique lower spots: cilia paler brown. Underside paler; band and spots as above; along exterior margin of forewing and exterior half of hindwing suffused with purple-grey. Antennæ brown, with a subapical white streak. Palpi and thorax beneath greyish-brown.

Expanse: $1\frac{4}{5}$ inches.

Habitat: Bengal." (*Moore, l. c.* in *P. Z. S.*)

" Larva pale green, white-speckled; head black-bordered. Feeds on *Zinziberaceæ*. Pupa pale green." (*Moore, l. c.* in *Lep. Cey.*)

Occurs Ceylon (*Hutchison*; *Wade*; *Mackwood*); Tavoy, Ponsekai (*Elwes* and *de Nicéville*); Andamans; Cachar (*Wood-Mason* and *de Nicéville*); Kumaon as *P. curvifascia* (*Doherty*); Orissa (*Taylor*); Sikkim (*de Nicéville*; *Elwes*); Nilgiris (*Hampson*); Himalayas, Assam, Ganjam, Wynaad, Travancore (*de Nicéville*).

Messrs. Wood-Mason and de Nicéville note that all their specimens from S. Andaman, the Sikkim hills, and the N.-E. frontier districts (Sibsagar, &c.) all agree with one another in always having three conjugated obliquely-placed subapical semi-transparent spots, and usually three in the reversed oblique series, the innermost of which is separated from the next to it by a greater interval than this is from the outermost, which latter is the absent one in those specimens with only two in the series. (*J. A. S. B., vol. l, pt. 2, p.* 256, *n.* 116 (1881))

Mr. de Nicéville notes that all the specimens of *P. alysos* he obtained in Sikkim during October had only one subapical white spot on the forewing (*J. A. S. B., vol. l, pt. 2, p.* 60, *n.* 128 (1881).

I have obtained numerous specimens of this species at Rangoon, about half of which have the innermost of the "reversely oblique" spots, all the other apical and reversely-oblique spots being wanting, the other half of my specimens have not even this one spot visible.

In collections Indian Museum and de Nicéville.

178.—NOTOCRYPTA PARALYSOS, WOOD-MASON and DE NICÉVILLE.

Plesioneura paralysos, Wood-Mason and de Nicéville, Proc. A. S. B., (1881), p. 143.

Plesioneura paralysos, Wood-Mason and de Nicéville, J. A. S. B., vol. l, pt. 2, p. 257, n. 117 (1881).

Notocrypta paralysos, de Nicéville, Journ. Bomb. Nat. Hist. Soc., vol. iv, p. 189, n. 3 (1889).

" ♂ and ♀. Closely allied to *P. alysos*, but differing therefrom, on the upperside of the anterior wings, in the discal oblique semi-transparent white lustrous band being broader with less irregular margins, and in only one small spot, placed between the third median and discoidal veinlet, midway between the discal band and the outer margin, being present; and, on the underside of the posterior pair, in having one or two small white opaque lustrous spots, one near the end of the cell, the larger and the more constantly present, and the other just beyond it between the first and second median veinlets.

Expanse: ♂ 1·66; ♀ 1·74 inches.

Habitat: S. Andamans.

Specimens from Sikkim hills, Sibsagar (*S. E. Peal*); Dhunsiri valley and the Dafla hills, Assam (*H. H. Godwin-Austen*); Trevandrum, S. India (*F. W. Bourdillon*); and Ceylon (*F. R. Mackwood*) are devoid of all traces of the spots on the lower surface of the posterior wings." (*Wood-Mason* and *de Nicéville, l. c.* in *J. A. S. B.*)

Mr. Elwes treats this species as synonymous with *N. alysos*. As Mr. de Nicéville considers the distinguishing character of this species to be the constant presence of a varying number of white opaque lustrous spots on the underside of hindwing in both sexes; it would appear that the specimens referred to above do not belong to this species, but that it is strictly confined to the Andaman Isles.

In collections Indian Museum and de Nicéville.

179.—NOTOCRYPTA ALBIFASCIA, *MOORE.*

Plesioneura albifascia, Moore, P. Z. S., 1878, p. 843, pl. liii, fig. 3, ♂.
Notocrypta albifascia, de Nicéville, Journ. Bomb. Nat. Hist. Soc., vol. iv, p. 190, n. 9 (1889).

" Male. Differs from *P. alysos*, Moore, in its narrower forewing and smaller hindwing ; the oblique semi-diaphanous white band is shorter and straight, the portion beneath the lower median branch being very small, this part on the underside showing only as a suffused white patch; and there are no subapical spots.

Expanse : 1½ inches.
Habitat : Hatsiega (Tenasserim).

A specimen of this species collected by Mr. Buxton, probably in Sumatra, is in my own collection." (*Moore, l. c.*)

In collection de Nicéville.

180.—NOTOCRYPTA RESTRICTA, *MOORE.*

Plesioneura restricta, Moore, Lep. Cey., vol. i, p. 178 (1881).
Notocrypta restricta, de Nicéville, Journ. Bomb. Nat. Hist. Soc., vol. iv, p. 189, n. 4 (1889).

" Allied to *P. alysos* : forewing with similar semi-transparent band, subapical spots, and also with a small linear spot between upper and middle median veins. Underside as above, with the band stopping at the subcostal vein in both sexes—not being continued in a white streak to the costal margin, as in *alysos*.

Expanse : 1¾ to 1⅞ inches." (*Moore, l. c.*)

Occurs in Ceylon (*Moore*) ; Orissa (*Taylor*) ; Cachar (*Wood-Mason* and *de Nicéville*) ; Sikkim (*de Nicéville* ; *Elwes*) ; Nilgiris (*Hampson*) ; Bhutan, Assam, Orissa (*de Nicéville*).

I have a single specimen of this species from Rangoon, and several from the Kadur District, Mysore.

In collections Indian Museum and de Nicéville.

181.—NOTOCRYPTA ASMARA, *BUTLER.*

Plesioneura asmara, Butler, Trans. Linn. Soc., Zool., second series, vol. i, p. 556, n. 3 (1877).
Hesperia asmara, Horsfield and Moore, MS., Cat. Lep. Mus. E. I. C., vol. i, p. 253, n. 576 (1857).
Plesioneura asmara, Distant, Rhop. Mal., p. 400, pl. xxxv, fig. 28 (1886).

" Similar to *P. dan*, but not tawny tinted, the three spots in centre of primaries united, and hyaline white.

Expanse : ♂ and ♀, 1 inch, 8 lines.
Habitat : Malay Peninsula ; Malacca (*Pinwill, Brit. Mus.*) ; Java (*Horsfield* and *Moore*)." (*Butler, l. c.*)

Mr. Distant quotes the above description and gives a figure from a Malaccan specimen in the British Museum. He also remarks that "this species is much more closely allied to *P. alysos*, and appears remote in appearance from *C. dan*."

I have numerous specimens of a *Notocrypta* from Rangoon which agree very closely with Distant's figure. I give below a fuller description than Mr. Butler's. Upperside brown, clothed with yellowish hairs at base of forewing and along inner margin of hindwing. Three transparent white spots in middle of forewing, the upper in cell largest, the middle one small and projecting beyond the other two, three small subapical white dots. Hindwing unmarked. Underside, spots on forewing as on upperside, two yellowish spots on costa above white band somewhat as in *N. alysos*, and a prominent diffused whitish spot at anal angle. Hindwing with the base clothed with yellowish hairs, causing the outer half to appear distinctly darker.

This species is evidently allied to *N. alysos*, but can be readily distinguished from it by the tone of the upperside which is brown and not nearly black, and also by the arrangement of the three discal spots on the forewing; the upper and lower ones are almost equal in size, the upper being slightly larger, and the middle one being a small triangular spot with its apex wedged between the other two, and its base projecting well beyond them. The upper and lower spots are connected internally for about half their width.

It also is apparently closely allied to *N. monteithi*, Wood-Mason and de Nicéville, with the figure of which it agrees fairly except that it entirely wants the fourth or lowest spot of the discal band, and also invariably has three white subapical dots.

It also appears to be very closely allied to the *C. cacus* and *C. consertus* of de Nicéville, especially to the former, but these species are described as belonging to the genus *Celænorrhinus* and so presumably rest with outspread wings, while the Rangoon species referred to above rests with wings raised above the back.

182.—NOTOCRYPTA MONTEITHI, *WOOD-MASON and DE NICÉVILLE*.

Plesioneura monteithi, Wood-Mason and de Nicéville, J. A. S. B., vol. lv, pt. 2, p. 391, n. 245, pl. xviii, figs. 3, 3 *a*, ♀ (1886).

Notocrypta monteithi, de Nicéville, Journ. Bomb. Nat. Hist. Soc., vol. iv, p. 190, n. 10 (1889).

" ♀. Upperside, both wings rich dark brown with a vinous tinge in some lights. Forewing with a semi-transparent white lustrous discal band of four completely-conjugated spots, the first large and oblong at the end of the cell; the second posterior and external to it at the base of the interspace between the second and third median nervules, and consequently wedge-shaped; the third conterminous with the first and second, forming an oblong figure whose opposite angles are subequal, and placed in the interspace between the first and second median nervules; and the fourth subpentagonal, double the size of the second, and half the size of the first and third, but not extending for more than two-thirds of the distance across the submedian area in which it is placed. Hindwing unmarked. Underside, forewing marked as above, but with the three subcostal streaks which connect the band with the costa much more distinct than above, where they are all but imperceptible; the lowest spot of the discal band much larger, reaching the submedian nervure, and outwardly diffused. Cilia concolorous with the wings.

Expanse: 1·7 inches.

Two females, Irangmara [Cachar], 8th and 18th July.

Closely allied to *P. feisthamelii*, Boisduval, from the Moluccas." (*Wood-Mason* and *de Nicéville, l. c.*)

In collection Indian Museum.

183.—NOTOCRYPTA BASIFLAVA, *DE NICÉVILLE.*

Plesioneura basiflava, de Nicéville, J. A. S. B., vol. lvii, pt. 2, p. 290, n. 22, pl. xiii, fig. 7 ♂ (1888).

Notocrypta basiflava, de Nicéville, Journ. Bomb. Nat. Hist. Soc., vol. iv, p. 193, n. 27 (1889).

" Male. Upperside, both wings dark glossy brown, with a slight vinous tinge. Cilia paler brown. Forewing with a pyramidal spot at the end of the cell, and a large somewhat rounded one below it in the first median interspace, both semi-transparent lustrous white. Hindwing unmarked. Underside, both wings rather paler than above. Forewing marked as on the upperside. Hindwing with the basal third of the wing rich chrome-yellow. Head and body above concolorous with the wings, palpi and thorax below grey, abdomen cinereous.

I am indebted to Mr. Harold S. Ferguson for a single specimen of this remarkable species, which, as far as I know, has no near ally. He informs me that it was captured by a Mr. Atholl MacGregor, probably in March or April, 1880, at Pirmaad, and that Mr. MacGregor, who is now in England, possesses only one other specimen.

Expanse: ♂, 1·8 inches.

Habitat: Travancore." (*de Nicéville, l. c.*)

Also recorded from the Nilgiris (*Hampson*).

In collection de Nicéville.

184.—NOTOCRYPTA BADIA, *HEWITSON.*

Pterygospidea badia, Hewitson, Ann. and Mag. Nat. Hist., fourth series, vol. xx, p. 322 (1877).

Pterygospidea badia, Hewitson, Desc. Lep. Coll. Atk., p. 4 (1879).

Plesioneura badia, de Nicéville, J. A. S. B., vol. lii, pt. 2, p. 88, n. 34, pl. x, fig. 10, ♂ (1883).

Notocrypta badia, de Nicéville, Journ. Bomb. Nat. Hist. Soc., vol. iv, p. 193, n. 28 (1889).

"Upperside dark brown. Anterior wing with a central narrow band, and four white spots near the apex (one of which is very minute and considerably below the rest) transparent white. Posterior wing with the fringe orange.

Underside as above, except that the posterior wing has a central yellow spot. Antennæ with a white ring near the point.

Expanse: $2\frac{1}{10}$ inches.

Habitat: Darjeeling (*Atkinson*)." (*Hewitson, Ann. Nat. Hist., l. c.*)

This seems to be a very rare species. Mr. de Nicéville states (J. A. S. B., 1883, p. 88) he has only seen two specimens, one in Col. Lang's collection, the other in the Indian Museum, Calcutta; both, as well as the specimen described by Hewitson, are from Sikkim. These two specimens have a fifth subapical small white spot, the extra one placed above the minute spot described by Hewitson in the lower discoidal interspace. The ring below the club of the antenna is ochreous, not white as stated by Hewitson·

In his list of Sikkim butterflies Elwes states that according to Möller it occurs at low elevations.

In collection Indian Museum.

The late Herr Plötz has also described two species of *Hesperiidæ* from "India" (*Plesioneura leucographa*, and *Plesioneura chimæra*) which in the absence of descriptions or specimens, Mr. de Nicéville has tentatively placed in his genus *Notocrypta*.

GENUS XLVI.—CELÆNORRHINUS.

Celænorrhinus, Hübner, Verz. bek. Schmett., p. 106 (1816).

Celænorrhinus, Plötz, Berl. Ent. Zeitsch., vol. xxvi, p. 253 (1882).

Gehlota, Doherty, Journ. A. S. B., vol. lviii, pt. 2, p. 131 (1889).

Celænorrhinus, de Nicéville, Journ. Bomb. Nat. Hist. Soc., vol. iv, p. 177 (1889).

Plesioneura (preoc.), part, auctorum.

"Forewing, *costa* slightly arched, *apex* rather acute, *outer margin* convex, *inner* margin straight; *costal nervure* terminating opposite the apex of the discoidal cell, *first, second*, and *third subcostal nervules* with their bases almost equi-distant, *fourth* subcostal with its base half as near to the base of the third subcostal as that vein is to the second, terminating at the apex of the wing, terminal portion of *subcostal nervure* or *fifth* subcostal nervule with its base almost touching that of the fourth, terminating on the outer margin far below the apex of the wing; *discoidal cell* long, narrow; *upper disco-cellular nervule* straight, strongly outwardly oblique, very short; *middle* and *lower* disco-cellular nervules almost in the same straight line (the lower a little concave), the lower a little longer than the upper, both veins taken together strongly inwardly oblique; *second median* nervule arising some little distance before the lower end of the cell, *first* median nervule arising much nearer to the base of the wing than to the point where the second median is given off; *submedian nervure* slightly recurved; *internal* nervure short and quickly running into the submedian nervure, with which it entirely anastomoses. Hindwing, *costa* strongly arched at base then straight to *apex*, which latter is somewhat acute in the male, rounded in the female, *outer margin* rounded, *inner* margin convex; *costal nervure* almost straight, terminating just before the apex of the wing; *first subcostal nervule* originating some distance before the apex of the cell; *upper disco-cellular* nervule straight, very slightly outwardly oblique; *lower* disco-cellular also slightly outwardly oblique, at first concave, then straight, a little longer than the upper disco-cellular; *discoidal* nervule very fine, straight, arising at the point of junction of the disco-cellular nervules; *second median* nervule arising just before the lower end of the cell, *first* median arising much nearer the lower end of the cell than the base of the wing; *submedian* and *internal nervures* straight. Type,* the *Papilio eligius* of Cramer.

This diagnosis has been made from bleached wings of both sexes of the "*Hesperia*" *leucocera* of Kollar, from Simla, and of the "*Papilio*" *eligius* of Cramer from the Amazons, for the specimens of which latter I am indebted to Dr. O. Staudinger. All the species of the genus settle with wide outspread wings, which at once distinguishes them in life from the genus *Notocrypta*, mihi, the species of which rest with wings folded upright over the back. *C. leucocera* in the Western Himalayas is markedly crepuscular, I have seen specimens over and over again flying up and down a short distance of the bed of the Simla river with immense rapidity, so fast that the eye can hardly follow them, settling on a leaf for a second and then flying off again, long after the sun has set. All that are known to me have the hindwing more or less spotted. *C. eligius*, Cramer, was described from Surinam in South America, and Felder states that he has received a specimen from Venezuela. The similarity in the markings of the forewing of this species to those of *C. maculosa*, Felder, from Shanghai, is not a little remarkable. The transformations of only one species are known, those of *C. spilothyrus*, Felder." (*de Nicéville, l. c.*)

185.—CELÆNORRHINUS PULOMAYA, *MOORE.*

Plesioneura pulomaya, Moore, P. Z. S., 1865, p. 787.

Celænorrhinus pulomaya, de Nicéville, Journ. Bomb. Nat. Hist. Soc., vol. iv, p. 180, n. 11 (1889).

---

* *Vide* Mr. Samuel H. Scudder in "Historical Sketch of the Generic Names proposed for Butterflies," Proc. Am. Acad. Arts and Sciences, vol x, p. 137 (1875).

"Male, dark olive-brown, yellowish-olive basally: forewing with four oblique discal semi-transparent white spots, the two upper large, one within and the other beneath the extremity of the cell, the third small and beneath the second, fourth also small and exterior to the juncture of the upper two: obliquely, before the apex, are five small similar spots, the upper three being conjugated; near posterior margin are two small orange-yellow spots, the first being one-third from the base, the other one-third from posterior angle: hindwing with three rows of irregular-shaped well-defined bright orange spots; cilia of hindwing broad, alternate brown and orange-yellow. Underside as above. Top of head black, with a marginal yellow narrow line on each side. Palpi above black, tipped with yellow. Palpi, thorax, and legs beneath yellow. Antennæ with yellow subapical streak.

Expanse: 2 inches.

Habitat: Darjeeling." (*Moore, l. c.*)

In recording this species from Kumaon Mr. Doherty states that he has obtained specimens transitional to *P. sumitra*, and considers it probably that the two are not specifically distinct.

Also recorded from Kangra, N.-W. Himalayas (*Moore*); Sikkim (*Elwes*).

According to Mr. Elwes this species is easily distinguished from other species of the genus. The antennæ are black, with a yellow or whitish band below the club; he also states that in Sikkim this species is restricted to the higher zones (7 to 10,000 ft.) where he never saw any other of the allied forms.

In collections Indian Museum and de Nicéville.

186.—CELÆNORRHINUS FLAVOCINCTA, *DE NICÉVILLE*.

Plesioneura flavocincta, de Nicéville, P. Z. S., 1887, p. 464, pl. xl, fig. 9, ♀.
Celænorrhinus flavocincta, de Nicéville, Journ. Bomb. Nat. Hist. Soc., vol. iv, p. 181, n. 12 (1889).

"Male. Upperside: both wings black. Forewing with the base (all except a round spot in the submedian interspace just beyond the origin of the first median nervule and touching it) thickly clothed with large tawny scales; a quadrate transverse spot beyond the middle of the cell, five conjoined subapical quadrate spots divided by the veins, the two lower ones shifted outwards, a narrow linear spot in the second median interspace, a quadrate one equal in size to that in the cell in the first median interspace, two much smaller ones placed obliquely (in one specimen conjoined in the right-hand wing) in the submedian interspace, all semi-diaphanous pale ochreous-white. Cilia black, all except a small portion on the submedian interspace, which is yellow. Hindwing with numerous more or less quadrate large orange spots disposed over the disc; the base of the wing clothed with long fur-like orange setæ. Cilia broadly orange, just marked with black at the end of the veins. Underside: forewing with the ground-colour paler, the spots as above, the inner margin below the submedian nervure ochreous, with two obscure ochreous diffused spots placed one above the other near the base of the wing in the submedian interspace. Hindwing with the colour of the ground apparently reversed, being orange, leaving a broad irregular black outer margin; a large black spot at the end of the cell, and a series of eight black spots placed one in each interspace (except the uppermost and lowest interspaces, which have two each) round the cell. Head black, marked with whitish at the base of the antennæ. Antennæ with the shaft above and club entirely, except the tip, creamy-white, the shaft below and tip of club black. Palpi whitish, marked anteriorly with three black lines, which meet at the apex. Thorax clothed above with long ferruginous setæ. Abdomen black, ringed with orange. Female. Differs only from the male in being larger.

Expanse: ♂ 2·3, ♀ 2·7 inches.

Habitat: Bhutan.

P. flavocincta is the largest species of the genus described from India. It is most nearly allied to P. pulomaya, Moore, and P. sumitra, Moore; but differs from both in the very large size of the orange spots on the hindwing on the upperside, these spots on the underside coalescing and occupying the greater portion of the wing, thus reducing the black ground-colour of the upperside to a band on the outer margin and to nine discal spots; in the two species just mentioned there is no tendency to this feature, the orange spots being all comparatively small and well-separated, and the base of the wing is black. The cilia in P. flavocincta are also very much broader, and orange throughout, except the bases of those cilia at the termination of the veins on the hindwing, which are black.

Described from a pair in the collection of Mr. A. V. Knyvett and a single male in that of Mr. Otto Möller, all of which were obtained near Buxa, Bhutan." (de Nicéville, l. c.)

In collection de Nicéville.

187.—CELÆNORRHINUS PYRRHA, *DE NICÉVILLE.*

Celænorrhinus pyrrha, de Nicéville, Journ. Bomb. Nat. Hist. Soc., vol. iv, p. 181, n. 17, pl. B, fig. 11, ♀ (1889).

" Habitat: Bhutan; Assam.

Expanse: ♂, 1·9; ♀, 2·0 to 2·2 inches.

Description: Male. Upperside, *forewing* dark brown, the basal half of the wing clothed with ochreous-yellow scales; a large square spot at the end of the discoidal cell, a rather smaller one below it in the first median interspace, a small one placed outwardly between these two spots in the second median interspace, two still smaller spots placed inwardly obliquely in the submedian interspace below the outer angle of the second spot, the lower one sometimes wanting, five small subapical spots arranged three and two—all these spots semi-transparent diaphanous white; *cilia* dark brown throughout. *Hindwing* dark brown, the basal two-thirds thickly clothed with long ochreous-yellow setæ, some bright yellow spots on the disc; *cilia* alternately dark brown and pale yellow. Underside, *forewing* spotted as above, but the anterior spot in the cell continued almost to the costa by two small white spots divided by the costal nervure, two diffused whitish spots placed in the submedian interspace beyond the two diaphanous spots of the upperside. *Hindwing* with all the spots more prominent and paler yellow than on the upperside. *Antennæ* below black throughout, above with a small portion just before the club shining silvery-white, the shaft dotted with white. Female: Upperside, *forewing* as in the male, but in some specimens there is a third white spot in the submedian interspace one-third from the base, and in some specimens also the spot in the cell has two small whitish dots above it almost reaching the costa. *Hindwing* as in the male, but the yellow spots more prominent. Underside, *forewing* as in the male, but with the pair of diffused whitish spots placed beyond the two oblique spots in the submedian interspace more prominent; *cilia* in this interspace often pale yellow. *Hindwing* as in the male. *Antennæ* as in the male.

The male differs from *C. sumitra*, Moore, from N.-E. Bengal (which is known to me by the description only), in having the shaft of the antennæ black anteriorly dotted with white, not anteriorly wholly silvery-white. The male differs from *C. pulomaya*, Moore, from Kulu, Sikkim, and Bhutan in having the lower of the two spots placed obliquely in the submedian interspace of the forewing in both sexes white, in *C. pulomaya* it is yellow. *C. putra*, Moore, from Bengal, is unknown to me; the description agrees, however, with some examples of *C. leucocera*, Kollar.

Described from a single male from Bhutan (I have examined the prehensores, so am certain that the specimen is a male), and six females also from Bhutan, one female from Cherrapunji, and one male and three females from the Khasi Hills." (*de Nicéville, l. c.*)

188.—CELÆNORRHINUS PLAGIFERA, *DE NICÉVILLE.*

Celænorrhinus plagifera, de Nicéville, Journ. Bomb. Nat. Hist. Soc., vol. iv, p. 182, n. 18, pl. B, fig. 13, ♀ (1889).

"Habitat: Sikkim, Bhutan.

Expanse: ♂, ♀, 2·0 to 2·3 inches.

Description: Male and Female. Upperside, *forewing* differs from *C. pyrrha*, mihi, in never having a spot one-third from the base in the submedian interspace. *Hindwing* with the spots larger, and of a richer (more orange) yellow colour; the alternate yellow portions of the *cilia* also of a deeper orange. Underside, *forewing* lacking the two diffused whitish spots in the submedian interspace beyond the two obliquely-placed transparent spots which are found in *C. pyrrha*; otherwise as in that species. *Antennæ* as in *C. pyrrha*.

I have described this species from nineteen specimens in the collections of Mr. A. V. Knyvett and myself. It appears to be very constant. The sexes are very difficult to discriminate; I have been able to distinguish them only by an examination of the organs of generation." (*de Nicéville, l. c.*)

189.—CELÆNORRHINUS PATULA, *DE NICEVILLE*.

Celænorrhinus patula, de Nicéville, Journ. Bomb. Nat. Hist. Soc., vol. iv, p. 182, n. 19, pl. B, fig. 4, ♀ (1889).

" Habitat : Sikkim.

Expanse: ♂, 2·2 ; ♀, 2·5 inches.

Description : Male. Upperside, *forewing* with the white spots forming the discal band smaller than in *C. pyrrha* and *C. plagifera*, mihi, no spot at the base of the second median interspace, the lower of the two spots in the submedian interspace minute. *Cilia* anteriorly dark brown, posteriorly pale yellow. *Hindwing* with the yellow spots on the disc larger and clearer than in either the above-mentioned species. *Cilia* almost entirely yellow, instead of being prominently marked with black at the ends of the veins. *Antennæ* with the shaft anteriorly entirely pure silvery-white, in which respect it agrees with *C. sumitra*, Moore, and *C. pero*, mihi. Female: Upperside, *forewing* with the lower spot in the submedian interspace larger than in the male, as also are the five subapical spots; a minute spot at the base of the second median interspace. *Cilia* posteriorly barely marked with pale yellow. *Antennæ* as in the male.

C. patula differs from the description of *C. sumitra* in having two spots in the submedian interspace of the forewing in both sexes instead of one only; the cilia are not alternately broadly brown and orange-yellow, and the female of *C. patula* lacks the yellow costal spot above the oblique discal series of white spots on the upperside of the forewing described in *C. sumitra*.

Described from a single pair from Sikkim. The female of *C. patula* is unique, as far as I know, amongst this group of the genus in possessing antennæ that are anteriorly white, this being usually a male character. I am certain of the sex of my type specimens, as I have examined the primary sexual organs." (*de Nicéville, l. c.*)

190.—CELÆNORRHINUS PERO, DE NICÉVILLE.

Celænorrhinus pero, de Nicéville, Journ. Bomb. Nat. Hist. Soc., vol. iv, p. 183, n. 20, pl. B., fig. 12, ♂ (1889).

" Habitat : India.

Expanse : ♂, 2·2 inches.

Description : Male. Upperside, *both wings* differ from those of *C. pyrrha*, mihi, in being paler in colour. *Forewing* marked as in that species, but the yellow spot one-third from the base of the wing in the submedian interspace more prominent ; *cilia* broadly pale yellow and brown. *Hindwing* as in *C. pyrrha*, but the yellow spots on the disc smaller, and the *cilia* almost entirely pale yellow, only just touched with brown at the ends of the veins, as in *C. patula* and *C. flavocincta*. Underside, *both wings* as in *C. pyrrha*, but the spot one-third from the base of the submedian interspace of the *forewing* much larger than on the upperside and white. *Antennæ* shining silvery-white anteriorly throughout, posteriorly black. *Palpi*, white below instead of pale yellow as in *C. pyrrha*.

C. pero agrees with *C. sumitra* in having the antennæ anteriorly white, but differs from the description of that species in having the palpi and front of thorax beneath white not pale yellow, and in possessing the additional yellow spot one-third from the base and the white spot one-third from the outer margin in the submedian interspace of the forewing on the upperside.

Described from two male examples from Colonel G. F. L. Marshall's collection. They are not ticketed, but I believe them to be from the Western Himalayas." (*de Nicéville, l. c.*)

Mr. de Nicéville has seen specimens of this species since the above description was written in Mr. H. J. Elwes' collection captured by Mr. W. Doherty in the Naga Hills.

191.—CELÆNORRHINUS SUMITRA, MOORE.

Plesioneura sumitra, Moore, P. Z. S., 1865, p. 787.

Celænorrhinus sumitra, de Nicéville, Journ. Bomb. Nat. Hist. Soc., vol. iv, p. 184, n. 17 (1889).

" Male and female dark olive-brown paler at the base. Male. Forewing with an oblique discal series of four conjugated semi-transparent white spots, the first and second large and quadrate, the other two very small ; first one within the extremity of the cell, second and third beneath it, fourth exteriorly before the juncture of the first and second ; a series of five small similar spots obliquely before the apex : hindwing with a submarginal row and a few discal bright orange-yellow spots ; cilia of hindwing broadly alternate brown and orange-yellow. Underside as above ; the orange-yellow spots on the hindwing more clearly defined,

Antennæ above silvery-white. Palpi and front of thorax beneath pale yellow. Abdomen with narrow orange-yellow segmental bands. Female, as in male, but having also a yellow costal spot above the oblique discal series of white spots.

Expanse: ♂ 2, ♀ 2⅜ inches.

Habitat: N.-E. Bengal." (*Moore, l. c.*)

Also recorded from Kumaon (*Doherty*); Sikkim (*Elwes*).

According to Mr. Elwes this species can be distinguished from *C. leucocera* by its large size, much larger and more numerous spots on the hindwing, and bright yellow, not greyish-yellow fringes; and can be distinguished from *C. pulomaya*, which it more nearly resembles in general appearance, by the colour of the antennæ, and the absence of the yellow spot on the hind-margin of the forewing.

Not in collection Indian Museum, unknown to de Nicéville.

Mr. de Nicéville considers that Mr. Elwes has probably wrongly identified his specimens. Rikisim. B. Busia. 5-7000 ft. Aug

192.—CELÆNORRHINUS LEUCOCERA, *KOLLAR*.

Hesperia leucocera, Kollar, in Hügel's Kaschmir, vol. iv, pt. 1, p. 454, n. 2, pl. xviii, figs. 3, 4, female (1848).

Celænorrhinus leucocera, de Nicéville, Journ. Bomb. Nat. Hist. Soc., vol. iv, p. 184, n. 21 (1889).

" All wings dark brown; some spots and a minute dot in the middle of the forewing, a small crescent at the apex, all transparent snow-white; on the upper surface of the hindwing four spots, on the under surface a larger number irregularly placed and yellowish, the borders white, interspersed with dusky.

Of median size, hindwing rounded at the outer margin, not elongated, the colour of all the wings above and below dark brown. In the middle of the forewing are three snow-white transparent almost quadrangular spots of which the hind one is the smallest, there is also a dot near the middle spot. Near the apex a crescent-shaped spot. On the hindwing several small yellow spots irregularly arranged; four on the upperside more or less obsolete, and seven on the underside almost punctiform, the outer five in a curved row near the outer margin. The fringe of the forewing brown with several whitish spots, on the hindwing alternately yellowish-white and brown. Head and body brown and hairy, the palpi yellowish-white clothed with brown hairs above; two yellow spots at the nape. Club of the antennæ snow-white, below brown.

Hügel brought a single specimen from the Himalayas.

Expanse: 1" [♀]" [♀, 2 inches from figure]. (*Kollar, l. c.*)

Also recorded from Bombay (*Swinhoe*); Orissa (*Taylor*); Ponsekai, Tavoy (*Elwes* and *de Nicéville*); Kumaon (*Doherty*); N.-W. Himalayas (*Moore*); North India (*Butler*); Sikkim (*Elwes*; *de Nicéville*).

This species is possibly conspecific with *P. munda*, as, according both to Mr. Elwes and Mr. Doherty, it varies in the number of spots on the hindwing which are sometimes entirely absent; Mr. de Nicéville, however, considers the two species to be distinct.

In collections Indian Museum and de Nicéville.

193.—CELÆNORRHINUS PUTRA, *MOORE*.

Plesioneura putra, Moore, P. Z. S., 1865, p. 788.

Celænorrhinus putra, de Nicéville, Journ. Bomb. Nat. Hist. Soc., vol. iv, p. 184, n. 20 (1889).

"Upperside dark fuliginous-brown; forewing with an oblique discal series of three conjugated semi-transparent whitish spots, the first and second large, the first within the extremity of the cell, the second beneath it, the third very small and at the lower outer angle of the latter; exterior to the junction of the upper two is a very minute spot, and before the apex is an interrupted series of five very small similar spots, the three upper being conjugated; hindwing with a submarginal and discal series of indistinct orange-yellow spots; forewing with ill-defined and hindwing with prominent, alternate yellowish-white and brown cilia. Underside dark brown, the semi-transparent spots whiter, and the submarginal and discal spots of the hindwing pale yellow. Palpi beneath pale yellow. Body and legs beneath brownish-yellow. Antennæ white in front.

Expanse: 2 inches.

Habitat: Bengal." (*Moore, l. c.*) Probably conspecific with the preceding.

Not in collection Indian Museum, unknown to de Nicéville.

194.—CELÆNORRHINUS MUNDA, *MOORE*.

Plesioneura munda, Moore, J. A. S. B., 1884, p. 48.

Celænorrhinus munda, de Nicéville, Journ. Bomb. Nat. Hist. Soc., vol. iv, p. 185, n. 23 (1889).

"Male and female. Upperside olive-brown: forewing with an oblique transverse discal semi-diaphanous white band, similar to, but more compact than in *P. leucocera*, the apex spot starting from above the costal vein, the two lower large spots, the small one beneath, as well as that outwardly between them, are not separated from each other, the second lower spot between the median and submedian only being apart from the rest; three subapical conjoined white spots and two minute lower dots; cilia very faintly alternated brownish-white; hindwing uniformly olive-brown: cilia deeply alternated with white. Underside paler than above; forewing marked the same: hindwing numerously speckled with olive-green scales towards abdominal margin; an olive-green speckled lunule at end of the cell.

Expanse: 1⅝ inches.

Habitat: Simla (*Lang*)." (*Moore, l. c.*)

Also recorded from Sikkim (*de Nicéville*).

In collection Indian Museum.

195.—CELÆNORRHINUS AMBAREESA, *MOORE.*

Plesioneura ambareesa, Moore, P. Z. S., 1865, p. 788.
Plesioneura ambareesa, de Nicéville, J. A. S. B., vol. lii, pt. 2, p. 87, n. 33, pl. x, fig. 9, ♀ (1883).
Celænorrhinus ambareesa, de Nicéville, Journ. Bomb. Nat. Hist. Soc., vol. iv, p. 185, n. 25 (1889).

" Upperside dark olive-brown, the whole surface irrorated with delicate yellowish-olive scales : forewing with an oblique transverse discal series of pale yellowish-white semi-transparent spots, the first large and within the extremity of the cell, being indented exteriorly, the second small and some distance beyond, the third large and beneath the latter ; below the last are two other small spots ; and one-third from the base beneath the cell is a small round similar spot ; above the first on the costa is a small brighter yellow spot, which is not transparent ; before the apex are five rather large similar spots, the three upper conjugated as are also the other two ; one or two lower submarginal very indistinct orange-coloured spots : hindwing with a row of ill-defined orange-coloured submarginal spots, and others still less defined nearer the base : cilia of both wings broad, alternate brown and yellowish-white. Underside paler, more uniform in colour ; markings as above.

Expanse : 2 inches.
Habitat : Bengal (Manbhoom).'' (*Moore, l. c.*)
Also recorded from Nilgiris (*Hampson*) ; Bombay (*Swinhoe*).
In collections Indian Museum and de Nicéville.

196.—CELÆNORRHINUS CHAMUNDA, *MOORE.*

Plesioneura chamunda, Moore, P. Z. S., 1865, p. 788.
Celænorrhinus chamunda, de Nicéville, Journ. Bomb. Nat. Hist. Soc., vol. iv, p. 185, n. 24 (1889).

" Female. Upperside dark brown ; base of forewing and the whole hindwing except exterior margin dark olive-brown ; forewing with an oblique transverse discal series of semi-transparent silky white spots, the first small and above the extremity of the cell, the second large and within the cell, the third beneath also large, the fourth facing the posterior angle, and a fifth exterior to and facing the juncture of the second and third ; a series of five small similar spots obliquely recurving before the apex ; cilia brown, on the hindwing alternated with white. Underside as above.

Expanse : 2 inches.
Habitat : Bengal." (*Moore, l. c.*) *to 4000ft. March to Nov.*
Also recorded from Sikkim by Mr. Elwes, who states it can be distinguished from *P. leucocera* by the colour of the hindwing, the antennæ also are never white as in *leucocera*, which sometimes, however, has only the club white.

In collections Indian Museum and de Nicéville.

197.—CELÆNORRHINUS NIGRICANS, *DE NICÉVILLE.*

Plesioneura nigricans, de Nicéville, J. A. S. B., vol. liv, pt. 2, p. 123, pl. ii, fig. 6, ♀ (1885).

Celænorrhinus nigricans, de Nicéville, Journ. Bomb. Nat. Hist. Soc., vol. iv, p. 186, n. 28 (1889).

" ♂ and ♀. Upperside swarthy, slightly sprinkled with ochreous scales on the basal half of the forewing. Forewing with an oblique semi-diaphanous pure white band across the disc composed of three conjoined spots, the middle one, at the base of the interspace between the first and second median nervules, the smallest, elongated, and projecting beyond the outer margin of the other two spots ; a small similar spot placed below the lower outer angle of the third spot, and an opaque spot placed above the uppermost spot not quite reaching the costa ; and with a recurved series of from three to five dots before the apex. Hindwing with the basal and abdominal areas sparsely clothed with long ochreous hairs. Underside, forewing as above, but the lowest spot of the discal band much larger, its inner margin straight, its outer margin blurred. Hindwing with an indistinct ochreous spot at the end of the cell, and a submarginal series of similar spots, the two towards the anal angle rather more distinct than the others. The base and abdominal area somewhat ochreous. Cilia dusky on the forewing excepting a small portion towards the inner angle which is ochreous : dusky also on the hindwing, with three ochreous spots below the outer angle, and one towards the anal angle. Antennæ blackish, the underside of the club and a small portion of adjoining shaft pure white.

Expanse : 1·7 inches.

Habitat : Sikkim (*Otto Möller*) ; Buxa, Bhutan (*Moti Ram*)." (*de Nicéville, l. c.*)

Also recorded from Tavoy (*Elwes* and *de Nicéville*), and Mr. de Nicéville informs me it is very common in the Khasi Hills.

In collections Indian Museum and de Nicéville.

198.—CELÆNORRHINUS FUSCA, *HAMPSON.*

Plesioneura fusca, Hampson, J. A. S. B., vol. lvii, pt. 2, p. 367, n. 261 (1888).

Celænorrhinus fusca, de Nicéville, Journ. Bomb. Nat. Hist. Soc., vol. iv, p. 186, n. 27 (1889).

" Differs from *P. spilothyrus* in having the cilia of the hindwing alternately black and white as in *P. leucocera* ; the costal bifid spot of the discal series on the forewing, white, not ochreous ; the underside mottled with obscure grey ; the latter half of the antennæ in the male white. The two lower spots of the subapical series on the forewing are often wanting, also the lowest spot of the discal series. Very near to *P. nigricans,* de Nicéville.

Expanse : 1·7 inches.

Habitat : Nilgiris and Shevaroy Hills [S. India] ." (*Hampson, l. c.*)

In collection de Nicéville.

199.—CELÆNORRHINUS SPILOTHYRUS, *FELDER.*

Eudamus spilothyrus, Felder, Verh. zool.-bot. Gesellsch. Wien, 1868, p. 283, ♂.

Plesioneura spilothyrus, Moore, Lep. Cey., vol. 1, p. 179, pl. 67, figs. 4, 4a (1881).

Celænorrhinus spilothyrus, de Nicéville, Journ. Bomb. Nat. Hist. Soc., vol. iv, p. 185, n. 26 (1889).

"Male and female. Upperside dark olive-brown; forewing with a few olive-yellow scales scattered on the basal area; a small spot on the middle of the costa, white in male, ochreous in female; a semi-diaphanous white quadrate spot with incised outer border at end of the cell; a similar-sized spot beneath it between the middle and lower median veins; a smaller spot above it at base of upper and middle medians, and one spot (in some specimens two) between the lower median and submedian near the posterior angle; a conjugated series of three small spots before the apex and two below them near the outer margin; hindwing with a very indistinct small spot at the end of the cell, and a curved discal series formed of clustered olive-yellow scales, these scales also extending over the base of the wing. Cilia cinereous-brown, in hindwing slightly alternated with olive-yellow. Underside paler; marked as above; the lower spots on forewing being olive-yellow. Body olive-brown; palpi with olive-yellow lateral streak. Antennæ with olive-white subapical streak.

Expanse: $1\frac{7}{10}$ to $1\frac{8}{10}$ inches." (*Moore, l. c.*)

Recorded from Ceylon (*Wade, Mackwood*); Nilgiris (*Hampson*).

In collections Indian Museum and de Nicéville.

200.—CELÆNORRHINUS CONSERTUS, *DE NICÉVILLE.*

Celænorrhinus consertus, de Nicéville, Journ. Bomb. Nat. Hist. Soc., vol. v, p. 222, n. 17, pl. E, fig. 12, ♂ (1890).

"Habitat: Khasi Hills.

Expanse: ♂, 1·7; ♀, 1·8 inches.

Description: Male. Upperside, *both wings* dark rich brown approaching to fuscous, the base of the forewing and the entire hindwing sprinkled with long ochreous-ferruginous hair-like scales. *Forewing* with three conjoined subapical lustrous white dots, the posterior rather nearer to the outer margin of the wing than the others; a compact discal lustrous white patch, anteriorly bounded by the subcostal nervure, posteriorly by the first median nervule, its inner edge nearly straight and even, its outer edge irregular owing to that portion of the patch situated at the base of the second median interspace being projected slightly forwards and beyond the line of the outer edge of the patch. *Hindwing* with a large rounded discoidal spot, with a similar discal series beyond formed by portions of the ground-colour being free from the ochreous-ferruginous hair-like scales. *Cilia* concolorous with the wings throughout. Underside, *forewing* entirely dark brown; the lustrous white spots as above, but the discal one with two pale yellow dots divided by the first subcostal nervule placed anteriorly against it; a subanal pale yellow patch divided into two by the submedian fold. *Hindwing* much as on the upperside. *Antennæ* black, with the shaft just below the club on the underside and the tip of the club beneath ochreous. Female. Upperside, *both wings* exactly as in the male, but the long hair-like scales less numerous. Underside, *both wings* as in the male.

Very probably near to "*Plesioneura*" *asmara*, Butler, from Malacca and Java, of which no sufficient description exists, but which Mr. Distant states is "closely allied to *P. alysos*," Moore, a species of the genus *Notocrypta*, in which I have provisionally placed "*P.*" *asmara*.

C. consertus has been described from two pairs obtained by the Revd. Walter A. Hamilton in the Khasi Hills." (*de Nicéville, l. c.*)

This species occurs in Rangoon as well as the succeeding one, and is not uncommon.

201.—CELÆNORRHINUS CACUS, DE NICÉVILLE.

Celænorrhinus cacus, de Nicéville, Journ. Bomb. Nat. Hist. Soc., vol. v, p. 223, n. 18, pl. E, fig. 11, ♂ (1890).

"Habitat: Rangoon.

Expanse: ♂, 1·6 inches.

Description: Male. Very near to *C. consertus*, mihi, but differing in the following particulars:—Smaller, wings narrower. Upperside, *both wings* much more thickly clothed with long ochreous hair-like scales than in that species, giving the ground a much more yellow appearance. *Forewing* with two subapical dots only, the lower most minute. Underside, *both wings* present the same differences as on the upperside.

Described from a single example taken at Rangoon in January, and deposited in the Phayre Museum." (*de Nicéville, l. c.*)

202.—CELÆNORRHINUS TABRICA, HEWITSON.

Pterygospidea tabrica, Hewitson, Ex. Butt., vol. v, pl. lix, fig. 8 (1873).

Celænorrhinus tabrica, de Nicéville, Journ. Bomb. Nat. Hist. Soc., vol. iv, p. 187, n. 35 (1889).

"Upperside dark brown. Anterior wing with eleven transparent spots: six in the centre, three of which are large, and five near the apex. Posterior wing orange, with the base and the costal and outer margins dark brown; a round black spot before the middle, and a submarginal band of seven oblong spots of the same colour. Underside as above, except that the anterior wing has a submarginal series of pale spots more distinctly marked as they approach the anal angle.

Expanse: $2\frac{1}{10}$ inches.
Habitat: Darjeeling (Roberts)." (*Hewitson, l. c.*)
This species is apparently closely allied to *C. pinwilli*, Butler.

203.—CELÆNORRHINUS PINWILLI, *BUTLER*.

Plesioneura pinwilli, Butler, Trans. Linn. Soc., Zool., second series, vol. i, p. 556, n. 5, pl. lxviii, fig. 4, ♂ (1877).

Plesioneura pinwilli, Distant, Rhop. Mal., p. 400, pl. xxxv, fig. 29, ♂ (1886).

Gehlota pinwilli, Doherty, J. A. S. B., vol. lviii, pt. 2, p. 131 (1889).

Celænorrhinus pinwilli, de Nicéville, Journ. Bomb. Nat. Hist. Soc., vol. iv, p. 187, n. 36 (1889).

The following is Mr. Butler's description as quoted in Rhop. Mal.:—

" Primaries black, with a bluish shot; a broad oblique shining pale yellow subhyaline patch, separated by the median nervure and its second and third branches into three spots; two small spots of the same colour, placed obliquely below it, on the interno-median interspace; secondaries bright orange, the base and apical portion of external border chocolate-brown; remainder of outer border, a rounded spot at end of cell, a second near anal angle, and five, submarginal, touching the outer border, black; head and thorax greenish-grey, vertex of head edged with sordid-white; abdomen orange, banded with black. Primaries below as above, excepting that there is a bifid whitish spot above the end of the cell, a whitish spot at base of interno-median area, and that the inner margin is brown; secondaries bright orange; the costal and outer borders irregularly purplish-black; fringe brownish; a subcostal dash, a rounded spot at the end of the cell, and a reniform spot near the anal angle black; body below and legs bright ochreous, palpi pale ochreous; neck below white; antennæ black above, testaceous below.

Expanse: 2 inches 2 lines.
Habitat: Malay Peninsula, Malacca (*Pinwill, Brit. Mus.*)."

Mr. Distant says that the single specimen captured by Capt. Pinwill is the only one known to him.

Mr. Doherty (*l. c.*) records one male from Margherita, Assam.

GENUS XLVII.—HANTANA.

Hantana, Moore, Lep. Cey., vol. i, p. 179 (1881).

" Differs from typical *Plesioneura* in the forewing being more regularly triangular, apex acute, exterior margin more oblique, and posterior margin shorter; cell much longer, extending to nearly three-fourths the wing, the disco-cellulars being very oblique; middle median at one-fifth and lower at four-fifths before end of the cell: hindwing regularly oval; cell longer; disco-cellulars more erect; medians nearer end of the cell. Palpi more laxly clothed; fore tibiæ tufted; middle and hind tibiæ hairy above, femora hairy beneath; antennal club shorter.

Type: *H. infernus*." (*Moore, l. c.*)

204.—HANTANA INFERNUS, *FELDER.*

Eudamus infernus, Felder, Verh. zool.-bot. Gesellsch. Wien, 1868, p. 283.
Hantana infernus, Moore, Lep. Cey., vol. i, p. 179, pl. 68, fig. 6 (1881).

" Male and female. Upperside dark purplish-brown; forewing with two, sometimes three, very small golden-yellow semi-diaphanous costal spots before the apex, a single but less distinct spot near the upper end of the cell, and in the female a similar costal spot above it. Underside slightly paler: golden-yellow spots on forewing brighter: hindwing with a very indistinct golden-yellow spot within the cell, one also at its end, and a series across the disc, all formed by golden-yellow scales. Head and palpi with lateral golden-yellow streak; front of head and palpi speckled with golden-yellow.

Expanse: ♂ $1\frac{5}{8}$; ♀ $1\frac{7}{8}$ inches." (*Moore, l. c.*)

Recorded from Ceylon (*Wade, Mackwood, Hutchison*).
In collections Indian Museum and de Nicéville.

GENUS XLVIII.—ASTICTOPTERUS.

Astictopterus, Felder, Wien. Ent. Monatsch., vol. iv, p. 401 (1860).
Astictopterus, Moore, Lep. Cey., vol. i, p. 162 (1881).
Astictopterus, Distant, Rhop. Mal., p. 401 (1886).

" Forewing elongated; exterior margin short, slightly oblique and convex, posterior margin long; cell broad, extending two-thirds the wing; first subcostal at nearly one-half before end of the cell, first, second, and third branches wide apart; disco-cellulars very slender and very oblique, upper bent inward, lower shortest; upper radial from the angle, lower from their middle; the middle median at one-sixth and lower at three-sixths before end of the cell; hindwing broadly oval; cell extending half the wing; disco-cellulars erect; radial very slender; second subcostal at one-sixth before end of the cell; middle median from close to end, and lower at nearly one half before end of cell. Body long, slender; palpi lax, terminal joint short, pointed; legs comparatively naked; antennæ with slender pointed club.

Type: *A. jama.*" (*Moore, l. c.*)

205.—ASTICTOPTERUS XANITES, *BUTLER.*

Astictopterus xanites, Butler, Trans. Ent. Soc. Lond., 1870, p. 510.
Astictopterus xanites, Distant, Rhop. Mal., p. 402, pl. 34, fig. 28 (1886).

" Wings above dark fuscous; anterior wings with a sanguineous transverse fascia not reaching costal or inner margin and crossing wing at end of cell. Wings beneath as above, but the transverse fascia to the anterior wings broader and paler than above and approaching nearer to the costal margin. Body and legs dark fuscous.

Expanse: wings, 32 millims." (*Distant, l. c.*)
Recorded from Tavoy (*Elwes* and *de Nicéville*).
I have obtained this commonly at Rangoon in August.
In collections Indian Museum and de Nicéville.

206.—ASTICTOPTERUS BUTLERI, *WOOD-MASON and DE NICÉVILLE.*

Astictopterus butleri, Wood-Mason and de Nicéville, J. A. S. B., vol. lii, pt. 2, p. 98, n. 260, pl. x, fig. 3 ♂ (1883).

Astictopterus butleri, Wood-Mason and de Nicéville, J. A. S. B., vol. lv, pt. 2, pt. 380, n. 205 (1886).

"♂. Closely allied to *A. xanites*, Butler, from the Malay Peninsula, Java, and Borneo. Upperside, both wings dark fuliginous-brown, strongly suffused with purple, and glossed with dark bronzy-brown according to the play of light. Forewing entirely lacking the broad orange band so conspicuous in *A. xanites*. Underside, forewing paler and less strongly suffused with purple, paling from the middle of the interno-median area to the posterior margin. Hindwing plum-coloured, with the anterior margin fringed at the base with black bristles forming a tuft, which is lodged in, and nipped by the lips of, an elongated sac-like fold in the cell of the forewing. ♀. Larger than the male. Upperside, forewing with an obsolete oblique discal red band, which is sometimes quite absent, but distinct traces of which are always to be seen on the underside, which is rather lighter everywhere and less stongly suffused with purple than in the male. The hindwing, of course, lacks the tuft of black bristles present in the opposite sex.

Expanse: ♂ 1·5; ♀ 1·7 inches.

Three males and four females in forests near Silcuri [Cachar], 8th to 19th July.

It is by no means an uncommon species at low elevations in Sikkim and Bhutan, and it has also been obtained in the Mergui Archipelago." (*Wood-Mason* and *de Nicéville, l. c.*, in *J. A. S. B.* of 1886).

Also recorded from Sikkim (*de Nicéville, Elwes*).

Near to *A. olivascens* but can be distinguished by the absence of markings on the underside of the hindwing.

In collections Indian Museum and de Nicéville.

207.—ASTICTOPTERUS OLIVASCENS, *MOORE.*

Astictopterus olivascens, Moore, P. Z. S., 1878, p. 692.

"Allied to *A. unicolor*, Bremer (Ménétriés, Cat. Mus. Pet., Lep., vol. i, t. 5, f. 6) from Pekin and Hong-kong but of larger size. Upperside uniform glossy olive-brown. Female with a white semi-diaphanous narrow subapical spot crossed by the fourth and fifth subcostal branches. Underside: male uniformly brown, forewing sparsely grey-speckled, hindwing with three ill-defined grey-speckled, transverse bands. Palpi, body, and legs beneath grey.

Expanse: $1\frac{4}{10}$ inches.

Habitat: Salween, Moulmein; Darjeeling." (*Moore, l. c.*)

Recorded from Calcutta (*de Nicéville*); Cachar (*Wood-Mason* and *de Nicéville*); Tavoy (*Elwes* and *de Nicéville*).

I have two specimens of this from Beeling, Upper Tenasserim. The shape of the wings is very similar to *Baracus*.

In collections Indian Museum and de Nicéville.

208.—ASTICTOPTERUS SALSALA, *MOORE*.

Nisionides salsala, Moore, P. Z. S., 1865, p. 786.

Astictopterus salsala, Distant, Rhop. Mal., p. 401, pl. xxxiv, fig. 21 (1886).

Astictopterus stellifer, Butler, Trans. Linn. Soc. Lond., Zoology, second series, vol. i, p. 455 (1876—79).

Astictopterus stellifer, Moore, Lep. Cey., vol. i, p. 163 (1881).

" Male and female dark brown with olive-brown gloss. Male; upperside, forewing with two or three ill-defined yellowish spots ascending obliquely from beyond middle of posterior margin. Female; forewing with an oblique series of small semi-transparent white spots curving across the disc (more or less distinct), and terminated below by an ill-defined yellowish spot. Underside chestnut-brown suffused with black on the disc; forewing with minute white spots, one at extremity of the cell, and two or three obliquely beyond; hindwing with a series of three spots disposed in a curve across disc; cilia greyish-brown. Palpi, body, and legs yellowish beneath.

Expanse: $1\frac{1}{8}$ inches.

Habitat: Bengal." (*Moore, l. c.*)

Also recorded from Cachar (*Wood-Mason* and *de Nicéville*); Tavoy (*Elwes* and *de Nicéville*); Calcutta (*de Nicéville*); Orissa (*Taylor*); Sikkim (*de Nicéville*; *Elwes*).

Recorded as *A. stellifer* from Ceylon (*Hutchison, Wade, Mackwood*); Poona, Bombay (*Swinhoe*); and the Nilgiris (*Hampson*).

Mr. de Nicéville states that he considers *A. salsala* to be identical with *A. stellifer*, though Mr. Moore informs him that the 'female of *A. salsala* has a curved discal row of seven white spots and two lower ochraceous discal spots, and is a larger species than *A. stellifer*, Butler.' According to Mr. Elwes the two species are identical, Sikkim specimens varying considerably in the spots of the forewing above, which are sometimes white, sometimes rufous and sometimes absent as in *stellifer*.

I have numerous specimens of this species from Rangoon, Beeling, Upper Tenasserim, Madras, Kadur District, and Mysore; they vary considerably in the distinctness of the spots both on upperside and underside, but I can find no sure characteristic by which to separate them into two species.

In collections Indian Museum and de Nicéville.

209.—ASTICTOPTERUS SUBFASCIATUS, *MOORE.*

Astictopterus subfasciatus, Moore, P. Z. S., 1878, p. 842.

"Male and female. Upperside uniform vinous-brown. Underside—forewing with a short vinous-grey streak, and two small ochreous-grey spots obliquely from the costa before the apex; hind-margin pale ochreous-brown; hindwing with a short, broad, medial, and a subbasal vinous-grey band; abdominal margin broadly pale ochreous-brown; some indistinct spots between the bands and at the base also pale ochreous-brown. Palpi, body, and legs beneath pale ochreous-brown.

Expanse: $1\frac{4}{8}$ inches.

Habitat: Ahsown (Upper Tenasserim).

A specimen of this species from S. India is also in the collection of the British Museum." (*Moore, l. c.*)

Recorded from Cachar (*Wood-Mason* and *de Nicéville*); Modah, Upper Burma (*Butler*); Nilgiris (*Hampson*). Messrs. Wood-Mason and de Nicéville note that "the male of this species has a curious impressed elongated oval brand on the forewing, placed so immediately behind as to touch the median nervure, and extending for one-third of its length along the first median nervule. This species probably has no real connection with the genus *Astictopterus*." (*J. A. S. B.*, vol. lv, pt. 2, p. 380, n. 204, pl. xviii, figs. 1, 1a, ♂ (1886).

Common throughout Burma in suitable localities; it seems to prefer deep shade.

In collections Indian Museum and de Nicéville.

GENUS XLIX.—KERANA.

Kerana, Distant, Rhop. Mal., p. 402 (1886).

"This genus is closely allied to *Astictopterus*, but structurally differs by having the first subcostal nervule of the anterior wings emitted more nearly opposite the base of the second than of the lower median nervule." (*Distant, l. c.*)

210.—KERANA DIOCLES, *MOORE.*

Nisionades diocles, Moore, P. Z. S., 1865, p. 787.

Kerana diocles, Distant, Rhop. Mal., p. 403, pl. 34, fig. 8 (1886).

"Male and female. Upperside uniform dark glossy olive-brown without markings. Underside pale brown, with a well-defined paler brown exterior border. Antennæ, palpi, and body dark olive-brown.

Expanse: ♂ $1\frac{3}{4}$; ♀ $2\frac{7}{8}$ inches.

Habitat: Bengal." (*Moore, l. c.*)

Recorded from Cachar (*Wood-Mason* and *de Nicéville*); Tavoy (*Elwes* and *de Nicéville*); Meetan, 3,000 ft., Upper Tenasserim (*Limborg*); Sikkim (*de Nicéville*; *Elwes*).

I have this species from Beeling, Upper Tenasserim.

In collections Indian Museum and de Nicéville.

211.—KERANA AURIVITTATA, *MOORE.*

Plesioneura aurivittata, Moore, P. Z. S., 1878, p. 843, pl. liii, f. 2.
Plesioneura aurivittata, var. cameroni, Distant, Rhop. Mal., p. 403, pl. 84, fig. 19 (1886).

" Male and female. Upperside dark golden olive-brown : forewing with a broad oblique golden-yellow discal band curving from the middle of the costa to posterior angle, the band semi-opaque from the costal vein to lower median branch : a small curved yellow streak composed of three vein-crossed spots before the apex. Cilia of both wings brown. Underside duller brown : forewing as above : hindwing slightly yellow-speckled, and with a very indistinct yellowish streak at the end of the cell.

Expanse : $1\frac{7}{10}$ inches.

Habitat : above Ahsown (Upper Tenasserim).

Allied to *P. dhanada,* Moore. Distinguished by the oblique transverse band being very broad throughout and extending to the posterior angle, also in the cilia of hindwing being entirely brown."* *(Moore, l. c.)*

Also recorded from Tavoy (*Elwes* and *de Nicéville*).

This species is readily distinguishable from *K. dhanada* by the extent of the yellow band on the forewing. The difference in the cilia does not hold good, that of *K. aurivittata* not being invariably brown on the hindwing.

I obtained a single specimen of this species at Beeling, Upper Tenasserim, and met with it very commonly at Tilin in the Yaw District, Upper Burma.

In collections Indian Museum and de Nicéville.

212.—KERANA GEMMIFER, *BUTLER.*

Astictopterus gemmifer, Butler, Trans. Linn. Soc., second series, Zoology, vol. i, p. 555 (1877).
Kerana gemmifer, Distant, Rhop. Mal., p. 403, pl. xxxiv, fig. 29 (1886).

" Wings above dark chocolate-brown ; anterior wings with a broad and outwardly rounded transversely oblique orange-yellow fascia crossing wings at about end of cell. Wings beneath as above, the fascia to anterior wings slightly paler. Body and legs concolorous with wings.

Expanse : 32 to 34 millim.

Habitat : Malay Peninsula ; Sungei Ujong (*Durnford—coll. Distant*) ; Malacca (*Pinwill—Brit. Mus.*) ; Singapore (*Wallace—coll. Godman* and *Salvin*)." (*Distant, l. c.*)

I have obtained two specimens of this species in the Hills about fifteen miles east of Toungoo, Burma, in dense, gloomy jungle. It is somewhat similar to *A. xanites* in appearance, but not only differs in neuration but is also a larger insect with a broader orange band.

Since writing the above Mr. de Nicéville has pointed out to me that my specimens, though possibly identical with Distant's species, are not the *K. gemmifer* of Butler, which has a number of amethystine spots on the underside of both wings, which are not present in the Toungoo specimens, and were also presumably wanting in Distant's specimens as he makes no mention of them: this last point is, however, doubtful as they are only visible in certain lights and may have escaped observation.

213.—KERANA DHANADA, *MOORE.*

Plesioneura dhanada, Moore, P. Z. S., 1865, p. 789.

"Upperside, dark yellowish olive-brown; the base of the wings brighter olive-brown: forewing with an oblique transverse discal irregular-margined semi-transparent yellowish band, joined above by a yellow costal spot; a small narrow streak of three conjugated similar spots obliquely before the apex: cilia brown, on the hindwing alternated with yellow. Underside; forewing as above, the lower portion of the oblique band terminating in a suffused yellow spot; hindwing with three transverse discal series of ill-defined yellowish-olive spots: cilia as above. Antennæ minutely spotted with yellow at the base, and with a subapical yellowish band. Palpi and thorax in front beneath yellow. Abdomen with narrow yellowish segmental bands.

Expanse: 1¾ inches.

Habitat: Bengal." (*Moore, l. c.*)

Recorded for Kumaon (*Doherty*); Sikkim (*de Nicéville*).

Mr. Elwes considers this species to be probably synonymous with *C. dan*, but has apparently never seen a specimen of it; while Mr. Doherty notes that he is not quite certain whether the two species belong even to the same genus; so apparently there is something wrong in the identification of it.

Mr. de Nicéville considers the two species certainly distinct, and informs me the present species should be included in *Kerana*.

In collections Indian Museum and de Nicéville.

GENUS L.—BARACUS.

Baracus, Moore, Lep. Cey., vol. i, p. 162 (1881).

"Wings small; forewing triangular; exterior margin short, convex, slightly oblique, posterior margin long; first subcostal at two-fifths before end of the cell, first, second, and third at equal distances, fourth and fifth much recurved from the base; disco-cellulars inwardly oblique, radial from their middle; cell extending beyond half the wing; middle median near to end and lower at nearly one-half before end of the cell; submedian straight; hindwing short, broadly oval; apex and exterior margin very convex; abdominal margin short; subcostal straight, second subcostal immediately before end of the cell; disco-cellulars slightly concave, radial from their middle; cell short; two upper medians from end of the cell, lower at one-third before the end; submedian and internal slightly recurved. Body moderate; palpi laxly clothed, terminal joint somewhat long, thick, pointed; hind tibiæ hairy above; antennæ with a thick club and pointed tip.

Type: *B. vittatus.*" (*Moore, l. c.*)

214.—BARACUS VITTATUS, *FELDER*.

Isoteinon vittatus, Felder, Verh. zool.-bot. Gesellsch. Wien, vol. xii, p. 480 (1862).

Baracus vittatus, Moore, Lep. Cey., vol. i, p. 162, pl. 69, figs. 1, 1 *a* (1881).

" Male and female. Upperside dark olive-brown. Male with the lower basal and discal area of both wings olive-grey, and a small subapical spot of the same colour also on the forewing. Female: forewing with a small olive-grey subapical spot and slender macular discal streak: hindwing with less distinct olive-grey lower basal and discal area. Underside ferruginous, the veins narrowly lined with paler ferruginous: forewing with the basal area dusky brown: hindwing with a longitudinal medial yellow fascia from base of cell, and less distinct short yellow discal streak between the veins. Body, palpi, and legs olive-brown, paler beneath.

Expanse: ♂ 1 $\frac{7}{12}$; ♀ $\frac{9}{12}$ inches." [♀ 1 $\frac{7}{12}$ inches?] (*Moore, l. c.*)

Habitat: Ceylon (*Hutchison, Wade, Mackwood*).

In collections Indian Museum and de Nicéville.

215.—BARACUS SUBDITUS, *MOORE*.

Baracus subditus, Moore, P. Z. S., 1883, p. 534.

" Female. Differs from the same sex of *B. vittatus* (*Isoteinon vittatus*, Felder) on the upperside, in being of a uniform olive-brown: forewing with three somewhat indistinct small olivaceous-yellow subapical spots, below which are four similar spots, the two lower of which are very indistinct. Hindwing uniformly olive-brown. Underside similar to that of *B. vittatus*, except that the forewing has no subapical or anal spots, and the intermediary streaks on the hindwing are more prominent.

Expanse: 1$\frac{1}{4}$ inches.

Habitat: Coonoor, Nilgiris (*Lindsay*)." (*Moore, l. c.*)

Also recorded from the Nilgiris by Mr. Hampson.

In collection de Nicéville.

216.—BARACUS SEPTENTRIONUM, *WOOD-MASON and DE NICÉVILLE*.

Baracus septentrionum, Wood-Mason and de Nicéville, J. A. S. B., vol. lv, pt. 2, p. 379, n. 203, pl. xviii, figs. 4, 4*a*, ♂ (1886).

" ♂. Upperside, both wings very dark vandyke-brown with golden, and in certain lights, greenish reflections. Forewing with an oblique subcostal streak, another shorter one below this in the cell, two subapical small spots separated from one another by the penultimate subcostal branch, and an indistinct spot on the disc between the second and third median nervules, opaque ochreous-brown. Underside, forewing greenish-black, increasingly bordered in front from the base to the apical angle, and thence decreasingly to the inner angle with ochreous-brown, varied by light streaks on the folds and by the two subapical spots of the upperside. Hindwing throughout ochreous-brown similarly varied with lighter streaks. Cilia smoky-brown in the forewing, paler on the hindwing.

Expanse: 1·5 inches.

One male, Irangmara [Cachar], 8th July, another very old specimen, also from Cachar, in the collection of the Indian Museum, Calcutta, and a third taken in the Sikkim Tarai by Mr. Otto Möller on 14th September, 1881.

B. septentrionum is nearly allied to *B. subditus*, Moore, but it differs therefrom in its considerably larger size, in the spots on the upperside of the forewing being of a brighter ochreous, in the shade of the ochreous of the underside generally, and in lacking the conspicuously-broad very pale yellow discal streak which extends from the base to near the outer margin of the hindwing and forms so marked a feature both in *B. subditus* from the Nilgiri and Pulni Hills and in *B. vittatus* (Felder) from Ceylon." (*Wood-Mason* and *de Nicéville, l. c.*).

Also recorded from the Nilgiris (*Hampson*).

I have a single specimen of this species from Beeling, Upper Tenasserim.

In collections Indian Museum and de Nicéville.

GENUS LI.—HESPERIA.

Hesperia, Fabricius, Ent. Syst., vol. iii, pt. i, p. 258, sec. 2, p. 325 (1793).

Pyrgus, Hübner, Verz. bek. Schmett., p. 109 (1816).

Scelothrix, Rambur, Catal. Lep. Andal., vol. i, p. 63 (1858).

Syrichthus, Boisduval, Icones, p. 230 (1832-33).

Hesperia, Moore, Lep. Cey., vol. i, p. 182 (1881).

"Wings small: forewing elongated, triangular; exterior margin short; cell of nearly uniform width and convex at the end; first subcostal beyond one-third before end of the cell, second and third branches at very wide and equal distances apart from the first; disco-cellulars convex, upper radial from angle near subcostal, lower from their middle; the middle median at one-sixth and lower at four-sixths before end of the cell; submedian straight; hindwing broadly conical; second subcostal at one-fourth before end of the cell, and in a line with base of the first; disco-cellulars slightly oblique and concave, slender; radial from their middle; the middle median from near end and lower about one-half before end of the cell. Body short; palpi lax in front, terminal joint stout, rather long and pointed; tibiæ pilose beneath, hind tibiæ with a long tuft of hair above; antennæ with a stout terminal club.

Type : *P. malvæ.*" (*Moore, l. c.*)

217.—HESPERIA DRAVIRA, *MOORE.*

Pyrgus dravira, Moore, P. Z. S., 1874, p. 576, pl. lxvii, fig. 5.

Pyrgus dravira, de Nicéville, Journ. A. S. B., vol. lii, pt. 2, p. 88, n. 36, pl. x, fig. 5, ♀ (1883).

" Allied to *P. marrubii.*

Female. Upperside dark greyish sap-brown, streaked with black between the veins. Cilia alternated with white. Forewing with a median triangular series of three diaphanous white spots, one being disposed at end of the cell and two on the disc; a geminated series of three smaller spots before the apex: hindwing with a prominent yellowish-white spot at end of the cell, and two smaller spots below it. Underside paler; forewing with markings as above: hindwing with greyish-white subbasal and discal spots, a streak from end of cell to outer margin, and a band along abdominal margin.

Expanse: 1⅜ inches.

Habitat: Cashmere (*Capt. H. B. Hellard*)." (*Moore, l. c.*)

Mr. Moore's figure of this species is not a good one, and does not agree with the description.

Recorded from Kandahar (*Swinhoe*).

In collections Indian Museum and de Nicéville.

This species and the next have nothing whatever to do with the typical species of the genus *Hesperia*, but should, as Mr. de Nicéville informs me, be placed in the genus *Erynnis*, Schrank.

218.—HESPERIA MARRUBII, *HERRICH-SCHÄFFER.*

Hesperia malvarum, var. *marrubii*, Herrich-Schäffer, Schmett. Eur., vol. i, *Hesp.* figs. 14, 15 (1845).

This is a species the correct identification of which is very doubtful; it has been confused with several other species of the same group, and will probably be found to be conspecific with some one of them. It is apparently very close to *P. dravira.* Mr. de Nicéville figures what he believes to be the latter species in J. A. S. B., 1883, pl. x, f. 5, ♀, but it is quite as likely to be the present species. In Kirby's Catalogue this species is placed as a synonym of *altheæ*, Hübner. Lang, in his European Butterflies. (p. 337, pl. 78, fig. 2), treats this species as synonymous with *bæticus*, Rambur, which latter he considers to be a variety of *P. altheæ*; the following is taken from his work quoted above:—

" *Spilothyrus altheæ*, Hübner, 452-3. *Malvarum* var., O., 1, 2, 107; Godt., ii, 28, 5, 6.

(S). *gemina*, Led., Z. B. V., 1852, p. 50.

(S). *marrubii*, Kirby, Man. Eur. Butt., p. 115.

Expands from 1·10 to 1·30 inches. Very closely resembles the last,* but is darker, and has a greenish tinge on the forewing. The wings are darker than in *alceæ*; the forewing has four white sub-diaphanous spots. Hindwing with two central white spots. Underside, hindwing tinged with green. The clubs of the antennæ are wholly black, whereas in the last species they are reddish-brown beneath.

Times of appearance—May and August.

Habitat : Central and South-Eastern Europe ; the larva of the type seems to be as yet unknown, though that of the Spanish variety *bæticus* is described by Rambur ; it may be inferred that there is some resemblance between the two forms.

VARIETY.

Bæticus, Rambur, Cat. Faun. And. P. C., 12, 3, 4 (1839); Rambur, Cat. S. And., p. 80.

Marrubii, Herrich-Schäffer, 14, 15.

Smaller than the type and lighter, being of a yellowish or brownish-grey colour.

Habitat : South-Western Europe.

Larva.—Pale grey, with a reddish or yellowish tinge, brown dorsal and lateral stripes. Feeds on *Marrubium Hispanicum* in April and August." (*Lang, l. c.*).

It will be seen that this description of the larva does not agree with that quoted below.

The larva of this species was met with by Major Roberts at Candahar in 1879 and is described as follows :—

" About 10''' long ; thickest in the middle, rather attenuated at each end. Skin soft, but with a ribbed and irregular surface, and covered with very short and minute whitish hairs. General colour dull (dusty) green ; dorsal line green, very fine and only visible on a few of the front segments. Head large, globular, slightly indented at the top, deep black (like charcoal), much larger than several segments which follow ; second segment smaller than head or third segment and forming a black neck or collar with three large yellow spots on it. Subdorsal stripe of a paler green than the ground-colour, but rather dull; spiracular, slightly raised or projecting flesh from the sides. Rokeran, Candahar, end of June ; wrapped up in the leaves of the mallow, on which it feeds.

Pupa wrapped up in a leaf, tightly webbed in and fastened by the tail only. Colour, brown washed with white." (*P. Z. S.*, 1880, p. 411, n. 23).

---

* *Spilothyrus alceæ*, Esper.

This species is recorded from N.-W. India (*Butler*); Quetta (*Swinhoe*); Kandahar (*Swinhoe*). I obtained what I believe to be this species at Quetta in June and July; my specimens agree very well with Mr. de Nicéville's figure referred to above, and not with Mr. Moore's figure of *P. dravira*. Mr. de Nicéville's figure is taken from a specimen obtained by him at Budrawar in Cashmere.

219.—HESPERIA CASHMIRENSIS, *MOORE*.

Pyrgus cashmirensis, Moore, P. Z. S., 1874, p. 274, n. 103, pl. xliii, fig. 7.

"Upperside dark fuliginous-black; body and base of wings with long grey hairs; cilia broad, alternate white and black; forewing with an irregular transverse discal series of eight small white spots, a white streak at end of cell and two narrow streaks above it: hindwing with three scarcely visible pale narrow discal streaks. Underside, greyish-brown, tinged with ochreous: forewing with spots as above; costa greyish-white: hindwing with anterior and abdominal margins grey; a white triangular subbasal spot, a broad transverse anterior discal patch with a small contiguous posterior spot, and a submarginal irregular series of spots.

Expanse: $1\frac{9}{10}$ inches.
Habitat: Cashmere." (*Moore, l. c.*)
Also recorded from Kumaon (*Doherty*).
In collections Indian Museum and de Nicéville.

220.—HESPERIA GALBA, *FABRICIUS*.

Hesperia galba, Fabricius, Ent. Syst., vol. iii, pt. 1, p. 352, n. 337 (1793).
Pyrgus galba, Butler, Cat. Fab. Lep., p. 281.
Pyrgus superna, Moore, P. Z. S., 1865, p. 792.
Hesperia galba, Moore, Lep. Cey., vol. i, p. 183, pl. 71, fig. 6 (1881).

"Upperside olive-brown: forewing with three pale yellowish-white spots within the cell, two beneath it, a transverse discal series of four spots followed by three minute subapical spots; a submarginal row of smaller spots: hindwing with a subbasal, a large medial, and small submarginal spots. Cilia of both wings alternate brown and pale yellow. Abdomen with narrow pale segmental bands. Underside paler olive-brown: forewing with the costal margin and spots as above, pale yellow: hindwing with a transverse subbasal, medial, and a narrow submarginal pale yellow maculated band. Palpi and body beneath, and legs, pale yellow.

Expanse: ♂ $\frac{3}{4}$; ♀ 1 inch." (*Moore, l. c.* in *Lep. Cey.*)

Occurs in Ceylon (*Wade, Mackwood, Hutchison*); Mhow, Bombay, Poona, Karachi (*Swinhoe*); Orissa (*Taylor*); Calcutta (*de Nicéville*); Kangra, N.-W. Himalayas (*Moore*); Kumaon (*Doherty*); Nilgiris (*Hampson*).

I have this species from Berhampore, Ganjam, and the Nilgiris.

I also have a single specimen from Poungadaw, Upper Burma, which Mr. de Nicéville considers belongs to this species. Indian specimens are considerably smaller and differ slightly in markings.

In collections Indian Museum and de Nicéville.

221.—HESPERIA EVANIDUS, *BUTLER.*

Pyrgus evanidus, Butler, Ann. and Mag. of Nat. Hist., ser. v, vol. v, p. 223, n. 14 (1880).

"Above extremely like *P. galba,* Fabricius (*P. superna,* Moore), but distinctly greyer in colour, the ground-colour being black instead of brown; below greyer and paler, the secondaries being very faintly tinted with yellowish, the central white belt broken up into three spots, of which the two lower ones are contiguous, instead of forming one continuous band across the wings; other markings similar.

Expanse of wings, 11 lines.

Habitat: Sao, Hubb River, Biluchistan, November." (*Butler, l. c.*)

Recorded from Karachi (*Swinhoe*); Campbellpore, N.-W. India (*Butler*). I found it not uncommon at Quetta.

In collection Indian Museum.

222.—HESPERIA ZEBRA, *BUTLER.*

Pyrgus zebra, Butler, Ann. and Mag. of Nat. Hist., ser. vi, vol. i, p. 207, n. 104 (1888).

Hesperia hellas, de Nicéville, Journ. Bomb. Nat. Hist. Soc., vol. iv, p. 177, n. 16, pl. B, fig. 9, ♂ (1889).

"Nearest to *P. sataspes* of South Africa; above black-brown; a spot in the cell, a smaller spot obliquely below it; a subtriangular spot across the end of the cell, two smaller spots obliquely below it, and a dot outside, forming a triangle with the disco-cellular and second spot; three small spots placed transversely between the subcostal branches halfway between the cell and apex; a curved series of five or six crescentic dots near to outer margin, a marginal series of dots at base of fringe, and a series of larger spots on the fringe white: secondaries with a subtrigonate spot at the end of the cell, a smaller oblong spot between the latter and the abdominal margin on the first median interspace, four or five dots near outer margin, a marginal series of spots, and the fringe white; palpi, edges of collar, and tegulæ greyish. Costal border of primaries below white; five black marginal dashes from the middle, the last dash being short and apical; a whitish patch at base of cell, a second at about centre of interno-median area, and a third at apex; the ordinary white spots larger than above; the fringe whitish, barred with blackish: secondaries greyish-brown; the base, an abbreviated, narrow, slightly zigzag, subbasal band, a broad, nearly regular band from costal to anal angle, and a narrow, slightly interrupted, stripe from apex to anal fourth of outer margin white; apical three-fourths of outer border grey; abdominal border white; fringe dull white, traversed by a greyish stripe: palpi excepting the tips, basal half of antennæ below, pectus, and legs white; venter white, the sides blackish with white edges to the segments.

Expanse: 26 millimetres.

Habitat : ♀ Campbellpore ; ♂ Futch Khan's Bungalow ; Kooteer, Chittar Pahar, 2,000 to 3,000 feet. April." (*Butler, l. c.*)

In describing this species from N.-W. India, Mr. Butler states that it is quite unlike any other Indian *Pyrgus*, the secondaries being alternately regularly banded with brown and white ; *P. evanidus*, with which it has been confused, differing from it in having the under surface of the hindwing olive-greenish, spotted and blotched with white. The species is described from specimens obtained by Major Yerbury in the above-noted localities.

Not in collection Indian Museum, but de Nicéville possesses two specimens from Campbellpore which he described as *H. hellas*.

GENUS LII.—LOBOCLA.

Lobocla, Moore, J. A. S. B., vol. liii, pt. 2, p. 51 (1884).

" Male. Forewing triangular, the edge of the costal margin slightly folded over on to the upperside from near the base to end of the costal vein*; the costal vein extending to three-fifths the margin ; subcostal five-branched, first branch emitted at one-third before end of cell, second and third at equal distances from the first ; fourth and fifth from end of the cell : disco-cellular bent outward near upper end and inwardly oblique hindward ; upper radial from the angle near subcostal, lower radial from the middle ; cell long, extending beyond two-thirds the wing ; three medians, lower at three-fourths and middle median at about one-fourth before end of the cell ; submedian straight ; hindwing short, broad, apex rounded, exterior margin slightly produced and angular at end of submedian vein ; costal vein extending to the apex ; subcostal touching the costal close to the base, two-branched, first branch at one-fourth before end of the cell ; disco-cellular very slender, almost erect ; the radial from its middle ; cell broad, extending to half the wing ; two upper medians from end of the cell, lower at about one-third before the end ; submedian and internal vein nearly straight. Body short, stout, thorax hairy ; palpi broad, thickly clothed, apical joint short, thick ; antennæ with a long slender-pointed tip ; femora and tibiæ short, stout, slightly pilose, middle tibiæ with two and hind with four spurs, tarsi long.

Type : *L. liliana*." (*Moore, l. c.*)

223.—LOBOCLA LILIANA, *ATKINSON*.

Plesioneura liliana, Atkinson, P. Z. S., 1871, p. 216, n. 3, pl. xii, fig. 2.

" Upperside dark fuliginous-brown. Forewing crossed from the middle of the costa to the posterior angle by a broad white semi-transparent band irregularly angled outwardly and narrowing from the middle hindwards. A recurved series of five angular white spots between the band and the anterior angle. Fringe brown, with a grey spot below the

---

* " The species of *Erynnis* (*E. alceæ*, &c.) have a similar fold on the costal margin of the forewing."

anterior angle, and two others near the posterior angle. Hindwing without markings, the fringe barred with greyish-white between the nervures. Underside:—Forewing with band and spots as above, but suffused with grey scales towards the apex, and pale towards the posterior angle below the band.

Hindwing suffused with greyish scales and crossed by irregular broken bands of grey, which give it a mottled appearance; two white dots near the base.

Palpi grey below, brown above.

Expanse of wings $2\frac{1}{4}$ inches.

Habitat: Yunan." (*Atkinson, l. c.*)

In collection de Nicéville, who informs me that it occurs commonly in the Khasi Hills.

224.—LOBOCLA CASYAPA, *MOORE.*

Lobocla casyapa, Moore, J. A. S. B., vol. liii, pt. 2, p. 52 (1884).

" Differs from *L. liliana* in its smaller size. Upperside somewhat paler and of an olive-brown tint, sparsely speckled with olive-grey scales: forewing with the transverse semi-diaphanous yellow band one-third less in width, the portions being distinctly defined by the traversing brown veins, the subapical spots also much smaller. Underside much paler; forewing numerously speckled with greyish-ochreous scales at the apex, the band and apical spots as above: hindwing with similarly disposed markings, but all composed of more numerous greyish-ochreous scales, these scales being whitish in *L. liliana*.

Expanse: $1\frac{9}{10}$ inches.

Habitat: Masuri (*Lang*), Cashmere (*Reed*)." (*Moore, l. c.*)

In collections Indian Museum and de Nicéville.

GENUS LIII.—GOMALIA.

Gomalia, Moore, P. Z. S., 1879, p. 144.

Gomalia, Moore, Lep. Cey., vol. i, p. 183 (1881).

" Wings short: forewing with the costa slightly arched at the base, apex acute, exterior margin oblique, posterior angle slightly convex; costal vein short; subcostal vein five-branched—first, second, and third arising before end of cell; fourth and fifth from its end; upper disco-cellular angled, lower oblique; upper radial from angle of upper disco-cellular, lower radial from its end; median vein three-branched, middle branch from near end of the cell; submedian vein nearly straight; hindwing lobed and angled near base of costal margin, apex and exterior margin very convex; costal vein extending to near apex, subcostal vein two-branched; one radial; median vein three-branched. Body short, thorax stout; palpi thickly pilose; antennæ short, with a thick very blunt club; legs moderately long, squamous." (*Moore, l. c.*)

225.—GOMALIA ALBOFASCIATA, *MOORE*.

Gomalia albofasciata, Moore, P. Z. S., 1879, p. 144.
Gomalia albofasciata, Moore, Lep. Cey., vol. i, p. 183, pl. 71, fig. 7 (1881).

" Upperside dark greyish-brown : forewing with a black transverse basal and discal band, a small white streak at end of the cell, two lunular spots on the disc, and three contiguous spots obliquely before the apex : hindwing with a broad white median transverse band. Underside paler, white markings as above. Palpi white beneath.

Expanse : 7/8 inch.

Habitat : Ceylon." (*Moore, l. c.*, in *P. Z. S.*)

Recorded from Kangra, N.-W. Himalayas (*Moore*) ; Nilgiris (*Hampson*).

I have obtained this species commonly at Ahmednagar, Deccan.

In collections Indian Museum and de Nicéville.

226.—GOMALIA LITORALIS, *SWINHOE*.

Gomalia litoralis, Swinhoe, P. Z. S., 1884, p. 513, n. 70, pl. xlvii, fig. 4.

" Karachi, July, 1879, in the salt-marshes on the seashore.

Allied to *G. albofasciata*, Moore. Larger, and more marked with white above ; costa arched, very nearly straight ; ground-colour similar. Forewing with a deep short white band occupying the space at the end of the cell, marked with black on the inner side, the black colour continued in the form of a band to the hinder margin, forming an elbow at the larger end of the white band ; a black band near the base, edged with whitish ; a lunular white spot on the disc, with a small white spot near it above ; a white streak running down the costa near the apex ; costa greyish ; fringe of the wing alternate brown and grey. Hindwing with a white spot at the base, a broad white discal band, and a deep white sinuous fringe. Below, the indications of the white markings are similar, but there is a white band at the base of the hindwing instead of a spot, and the entire surface of both wings is of a suffused pale bronzy-brown colour, with all the markings suffused and indistinct." (*Swinhoe, l.c.*)

Not in collection Indian Museum, unknown to de Nicéville.

SPECIES INCERTÆ SEDIS.

I am unable to discover to what genera the following four species really belong, so have left them undecided in the genera *Pamphila* and *Hesperia* in which they were originally described, all except the last which Hewitson placed in the South American genus *Eudamus*.

227.—PAMPHILA AVANTI, *DE NICÉVILLE*.

Pamphila avanti, de Nicéville, J. A. S. B., vol. lv, pt. 2, p. 255, n. 9, pl. xi, fig. 10, ♂ (1886).

" Male. Upperside, both wings fuscous. Forewing with a broad oblique subbasal streak, a more irregular and broader discoidal streak, both commencing close to the costa and extending to the submedian nervure when they are joined, and a subapical streak, which joins the discal one, all yellow. Cilia very long and brown. Hindwing with a large irregular-shaped spot in the middle of the disc, with two very small and indistinct ones placed outwardly beyond it. Cilia pale yellow. Underside, both wings much paler, ferruginous-ochreous on the hindwing; a dark anteciliary line. Forewing with the yellow markings much as above but paler and more extended. Hindwing with the discal spot large and silvery-ochreous, with a small spot in the cell near the base, and another of the same size beyond the outer end of the large discal spot, a lengthened pale ochreous abdominal streak. Antennæ short, with a prominent club, the shaft and club fuscous above, the club tipped with ferruginous below, and the shaft pale ochreous.

Expanse: ♂ 1 inch.

Habitat: Native Sikkim.

There are two specimens of this very pretty and distinct species in Mr. Otto Möller's collection, obtained by his Native Collectors probably at high elevations in Sikkim near the passes. It is unlike any hesperid known to me." (*de Nicéville, l. c.*)

Types in collections Möller and Elwes.

228.—PAMPHILA DIMILA, *MOORE*.

Pamphila dimila, Moore, P. Z. S., 1874, p. 576.

" Allied to *P. comma*.

Male and female. Upperside testaceous; exterior border broadly fuliginous-brown; apex of forewing brownish-testaceous. Cilia whitish-testaceous: forewing with a series of small yellow apical spots; male with an oblique silvery-lined black streak below the cell: hindwing with a yellow spot within the cell, and a curved discal series of four quadrate spots. Underside: forewing pale testaceous; apical spots as above: hindwing with basal portion greenish-brown; three prominent white subbasal spots disposed above, below, and at end of the cell; a curved discal series of six quadrate white spots.

Expanse: ♂ 1⅜; ♀ 1¾ inches.

Habitat: Runang Pass, Busahir (S.-E. side, about 13,000 feet). (*Capt. H. B. Hellard*)." (*Moore, l. c.*)

229.—HESPERIA CYRINA, *HEWITSON*.

Hesperia cyrina, Hewitson, Ann. and Mag. of Nat. Hist., ser. iv, vol. xviii, p. 450 (1876).

Parnara parca, de Nicéville, Journ. Bomb. Nat. Hist. Soc., vol. iv, p. 174, n. 13, pl. B, fig. 10, *female* (1889).

Parnara parca, Watson, *Hesp. Ind.*, p. 49, n. 65 (1891).

"Upperside dark brown. Antennæ ringed with white near the point. Anterior wing with seven transparent spots—one in the cell bifid, three placed longitudinally below this, and three near the apex. Posterior wings with the five transparent spots forming a circle: the fringe, except at the apex, orange. Underside as above, except that the anterior wing has the outer margin near the anal angle yellow, and that the posterior wing has the outer margin between the branches of the median nervures, as well as the fringe, yellow and indented inwardly.

Expanse: 1·8 inches.

Habitat: Darjeeling.

Unlike any described species." (*Hewitson, l. c.*)

In collection de Nicéville.

230.—HESPERIA DECORATUS *HEWITSON*.

Eudamus decoratus, Hewitson, Desc. Hesp., p. 17, n. 30 (1867).

Pterygospidea decoratus, id., Ex. Butt., vol. v, pl. *Pterygospidea*, fig. 2 (1873).

"Upperside orange, with bands and spots of black; the abdomen orange banded with black. Anterior wing with two spots near the base, a band of three spots before the middle, a triangular spot on the costal margin beyond the middle and a spot below this, all black; the apex and outer margin broadly brown crossed longitudinally by fine hastate yellow lines. Posterior wing with a spot at the base, two spots before the middle, and two submarginal bands of pyriform spots near the apex. Underside as above, except that the black spots are larger, and that the anal angle of the posterior wing is black. A specimen of this species in the collection of Dr. Boisduval is white instead of orange. (*Hewitson, l. c.,* in *Ex. Butt.*)

There is not in the whole family of the *Hesperiidæ* a more beautiful species than this." (*Hew., l. c.,* in *Desc. Hesp.*)

Expanse: 1·6 inches.

Habitat: Sylhet, Java (*Hewitson*); Garo Hills (*de Nicéville*).

In collection de Nicéville.

INDEX.

(SYNONYMS ARE PRINTED IN *Italics*, GENERA IN CAPITALS.)

A

	Page
ABARATHA	98
Acroleuca, Erionota	107
Aditus, Suastus	52
Adrastus, Hyarotis	117
AEROMACHUS	65
Agama, Pyrgus	100
Agna, Chapra	32
Agni, Tapena	122
Aina, Halpe	72
Aitchisoni, Pithauriopsis	28
Albicilia, Sarangesa	55
Albifascia, Notocrypta	128
Albofasciata, Gomalia	159
Alexis, Parata	17
Alica, Tagiades	94
Alysos, Notocrypta	126
Amara, Choaspes	8
Ambareesa, Celænorrhinus	140
AMPITTIA	61
Anadi, Choaspes	7
Anura, Hasora	12
Aria, Matapa	22
Asmara, Notocrypta	128
Assamensis, Parnara	37
ASTICTOPTERUS	145
Atkinsoni, Isoteinon	77
Atticus, Tagiades	95
Attina, Unkana	4
Augias, Telicota	55
Aurivittata, Kerana	149
Austeni, Parnara	43
Avanti, Pamphila	159

B

	Page
Bada, Hesperia	34
BADAMIA	3
Badia, Notocrypta	130
Badra, Hasora	12
Bambusæ, Telicota	56
BAORIS	29
BARACUS	150
Basiflava, Notocrypta	130
Benjamini, Choaspes	5
Beturia, Halpe	70
Bevani, Parnara	36
Bhagava, Satarupa	88
Bhawani, Hidari	112
BIBASIS	15
Brahma, Telicota	57
Brunnea, Halpe	76
Buchananii, Tapena	124
Butleri, Astictopterus	146

C

	Page
Cacus, Celænorrhinus	143
Cahira, Parnara	43
CALLIANA	91
Callineura, Plastingia	113
Camertes, Cyclopides	61
Canaraica, Parnara	42
CAPILA	25
Cashmiriensis, Hesperia	155
CASYAPA	108
Casyapa, Lobocla	158
CELÆNORRHINUS	131
Cephala, Isoteinon	80
Cephaloides, Isoteinon	80
Ceramas, Taractrocera	63
Cerata, Halpe	73
Ceylonica, Halpe	76
CHÆTOCNEME	108
Chamunda, Celænorrhinus	140
CHAPRA	31
Chaya, Hesperia	32
CHOASPES	5
Chromus, Parata	16
Cicero, Hesperia	125
Cingala, Parnara	35
Colaca, Parnara	36
COLADENIA	118
Consertus, Celænorrhinus	142
Coras, Ampittia	61
Coulteri, Hasora	14
Crawfurdi, Choaspes	9
CTENOPTILUM	101
CUPITHA	64
CYCLOPIDES	68
Cyrina, Hesperia	160

D

	Page
Dan, Coladenia	120
Danna, Taractrocera	63
Dara, Padraona	57
DARPA	106
Dasahara, Sarangesa	54
Decorata, Halpe	76
Decoratus, Hesperia	161
Dhanada, Kerana	150
Dimila, Pamphila	160
Diocles, Kerana	148
Distans, Tagiades	93
Divodasa, Hesperia	51
Dolopia, Halpe	74
Dravira, Hesperia	153
Druna, Matapa	23

E

	Page
Eitola, Parnara	45
ERIONOTA	107
Evanidus, Hesperia	156
Exclamationis, Badamia	3

F

Fatih, Coladenia	119
Farri, Parnara	44
Flaccus, Hesperia	62
Flavalum, Isoteinon	83
Flavipennis, Isoteinon	81
Flavocincta, Celænorrhinus	133
Flexilis, Isoteinon	85
Folus, Udaspes	125
Fusca, Celænorrhinus	141

G

Galba, Hesperia	155
Gana, Tagiades	96
GANGARA	110
GEHLOTA	131
Gemmifer, Kerana	149
Glandulosa, Paduka	19
Gola, Padraona	59
Goloides, Padraona	60
GOMALIA	158
Gomata, Choaspes	7
Gopala, Satarupa	90
Gremius, Suastus	51
Gupta, Halpe	73
Guttata, Parnara	34

H

Hadria, Hasora	13
HALPE	70
Hamiltonii, Coladenia	121
Hanria, Darpa	106
HANTANA	144
Harisa, Choaspes	6
HASORA	11
Helferi, Tagiades	98
Hellas, Hesperia	156
HESPERIA	152
HIDARI	111
Hiraca, Hesperia	107
Honorei, Halpe	75
HYAROTIS	117
Hypapa, Erionota	112

I

Lapis, Isoteinon	86
Indistincta, Aëromachus	66
Indrani, Coladenia	118
Indrasana, Isoteinon	86
Infernus, Hantana	145
Irava, Hidari	112
ISMENE	9
ISOTEINON	77

J

Jaina, Ismene	10
Jayadeva, Capila	25
Jhora, Aëromachus	68
Julianus, Hesperia	31

K

	Page
Kali, Aëromachus	67
Karsana, Chapra	34
KERANA	148
Khasiana, Tagiades	92
Khasianus, Isoteinon	78
Kumara, Halpe	72
Kumara, Parnara	41

L

Ladon, Papilio	3
Lalita, Erionota	109
Latoia, Hesperia	113
Latreillei, Hesperia	4
Laxmi, Tapena	123
Lebadea, Paduka	19
Leucocera, Celænorrhinus	138
Lidderdalii, Casyapa	109
Liliana, Lobocla	157
Litoralis, Gomalia	159
LOBOCLA	157

M

Mæsa, Pamphila	57
Mæsoides, Padraona	58
Mævius, Taractrocera	62
Mahintha, Ismene	11
Malayana, Parata	18
Mangala, Pamphila	34
Margherita, Plastingia	115
Maro, Ampittia	61
Marrubii, Hesperia	153
Masoni, Isoteinon	85
Masuriensis, Isoteinon	79
MATAPA	21
Mathias, Chapra	31
Meetana, Tagiades	94
Menaka, Tagiades	95
Microstictum, Isoteinon	82
Minuta, Tagiades	96
Modesta, Isoteinon	84
Mölleri, Suastus	53
Monteithi, Notocrypta	129
Moolata, Parnara	42
Multiguttata, Ctenoptilum	103
Munda, Celænorrhinus	139
Murdava, Pithauria	27

N

Naga, Plastingia	115
Narada, Satarupa	89
Narooa, Parnara	39
Nigricans, Celænorrhinus	141
Nilgiriana, Isoteinon	84
Noëmi, Plastingia	114
Nostrodamus, Chapra	33
NOTOCRYPTA	125

O

Obscurus, Tagiades	93
Obsoleta, Aëromachus	67
Oceia, Baoris	29
ODONTOPTILUM	104
Œdipodea, Ismene	10
Olivascens, Astictopterus	146
Ornata, Parnara	38

INDEX.

P

	Page
PADRAONA	57
PADUKA	18
Pagana, Parnara	40
Palmarum, Padraona	60
Pandita, Isoteinon	81
Pandia, Hesperia	110
Paralysos, Notocrypta	127
PARATA	16
Parca, Parnara	49, 160
PARNARA	34
Patula, Celænorrhinus	136
Penicillata, Baoris	30
Pero, Celænorrhinus	137
Phanæus, Casyapa	109
Phænicis, Hesperia	117
Phisara, Satarupa	89
Pholus, Parnara	47
Pieridoides, Calliana	91
Pinwilli, Celænorrhinus	144
PIRDANA	20
PISOLA	26
PITHAURIA	26
PITHAURIOPSIS	28
Plagifera, Celænorrhinus	135
PLASTINGIA	113
Plebeia, Parnara	40
PLESIONEURA	125
Potiphera, Pterygospidea	99
Praba, Plesioneura	117
Pralaya, Tagiades	97
Prominens, Chapra	33
Pseudomæsa, Padraona	59
PTERYGOSPIDEA	91
Pulomaya, Celænorrhinus	132
Purendra, Sarangesa	54
Purreea, Cupitha	64
Putra, Celænorrhinus	139
PYRGUS	152
Pyrrha, Celænorrhinus	134

R

	Page
Radians, Halpe	74
Ransonnetii, Abaratha	99
Ravi, Tagiades	92
Restricta, Notocrypta	128
Rudolphei, Pirdana	20

S

	Page
Sagara, Pamphila	62
Sala, Suastus	51
Salsala, Astictopterus	147
Sambara, Satarupa	89
Sarala, Parnara	48
SARANGESA	53
Saraya, Abaratha	99
Sasivarna, Matapa	23
SATARUPA	87
Satwa, Isoteinon	79
SCELOTHRIX	152
Scopulifera, Baoris	29
Semamora, Parnara	46
Sena, Bibasis	15
Separata, Halpe	71
Septentrionum, Baracus	151
Seriata, Parnara	42
Shalgrama, Matapa	24
Sikkima, Halpe	70

	Page
Sitala, Halpe	75
Siva, Telicota	57
Spilothyrus, Celænorrhinus	142
Stellifer, Astictopterus	147
Stigmata, Aëromachus	68
Stramineipennis, Pithauria	27
SUASTUS	50
Subditus, Baracus	151
Subfasciata, Matapa	24
Subfasciatus, Astictopterus	148
Subgrisea, Suastus	53
Subochracea, Chapra	32
Subradiatus, Cyclopides	69
Subtestaceus, Isoteinon	78
Subvittatus, Cyclopides	69
Sumitra, Celænorrhinus	137
Superna, Pyrgus	155
Sura, Odontoptilum	105
Swerga, Suastus	53
SYRICHTHUS	152
Syrichthus, Abaratha	100

T

	Page
Tabrica, Celænorrhinus	143
TAGIADES	91
TAPENA	121
TARACTROCERA	62
Taylorii, Abaratha	99
TELICOTA	55
THANAOS	65
Thrax, Erionota	107
Thwaitesi, Tapena	122
Thyrsis, Gangara	110
Tissa, Coladenia	118
Toona, Parnara	45
Trichoneura, Tagiades	97
Tulsi, Parnara	44
Tympanifera, Cupitha	64

U

	Page
UDASPES	124
Uma, Parnara	38
Unicolor, Baoris	29
UNKANA	4

V

	Page
Vasava, Ctenoptilum	103
Vasutana, Choaspes	8
Vindhiana, Isoteinon	84
Vitta, Hasora	14
Vittatus, Baracus	151

W

	Page
Watsonii, Parnara	46

X

	Page
Xanites, Astictopterus	145
Xanthopogon, Hesperia	5

Z

	Page
Zebra, Hesperia	156
Zema, Halpe	74
Zennara, Pisola	26

www.ingramcontent.com/pod-product-compliance
Lightning Source LLC
Chambersburg PA
CBHW020258170426
43202CB00008B/430